Diversity and Disagreement

Adam Feltz • Edward T. Cokely

Diversity and Disagreement

From Fundamental Biases to Ethical Interactions

palgrave
macmillan

Adam Feltz
Department of Psychology
The University of Oklahoma
Norman, OK, USA

Edward T. Cokely
Department of Psychology
The University of Oklahoma
Norman, OK, USA

ISBN 978-3-031-61934-2 ISBN 978-3-031-61935-9 (eBook)
https://doi.org/10.1007/978-3-031-61935-9

This Palgrave Macmillan imprint is published by the registered company Springer Nature Switzerland AG.
The registered company address is: Gewerbestrasse 11, 6330 Cham, Switzerland

Paper in this product is recyclable.

To Silke, who has made me more ethical.
To Colleen, for giving me a chance to learn when no one else would.

Acknowledgments

The core ideas in this book started with a random (or perhaps fated? or determined?) encounter about 15 years ago. Two graduate students from two different departments were attending an experimental economics reading group hosted by a third department when they struck up a conversation about individual differences. What happened from there is documented in this book.

We've had indescribable support and encouragement from numerous people—too many to thank in a short acknowledgment. But we'd like to thank Alfred Mele, Michael Bishop, Mark Isaac, David McNaughton, K. Anders Ericsson, Paul Ward, Lael Schooler, Neil Charness, Gerd Gigerenzer, and Colleen M. Kelley for being supportive of our blossoming vision as graduate students and recent graduates. We'd like to thank the experimental philosophy community for being such an encouraging, and constructive group. We'd like to thank the people of Schreiner University, Michigan Technological University, the Center for Adaptive Behavior and Cognition at the Max Planck Institute for Human Development, and the University of Oklahoma for providing environments to help develop this work. We offer special thanks to our colleague and friend Rocio Garcia-Retamero for offering so much support along the way. We are grateful to Scott Gronlund, Kelvin Droegemeier, Jorge Mendoza, Shane Connelly, Eric Day, Carol Silva, Hank Jenkins-Smith, and several others who made it possible for us to join the faculty of The University of Oklahoma. We'd also like to acknowledge generous financial support provided by the Templeton Foundation; the National Science Foundation; The U.S. Geological Survey; the National Academies of Science, Engineering, and Medicine; The United States Department of Commerce and National Oceanic and Atmospheric Administration; The Spanish Ministry of Science & Innovation; Medscape Education; The Arete Initiative at the University of Chicago; Time Sharing Experiments in the Social Sciences; The University of Oklahoma Transportation Center; and the National Institute for Risk and Resilience at The University of Oklahoma.

We'd like to thank the numerous students, postdocs, and early-career scholars who have in one way or another provided feedback, criticism, and ideas for this book, including (but not limited to) Braden Tanner, Uyen Hoang, Dana Mahmoud-Elhaj, Mohammad Asif, Jenna Holt, Stephanie Samayoa, Megan Harris, Ashley Perez, Melissa Millan, Taylor Abt, Tom Offer-Westort, Eric Schulz, Paula Parpart, Stephanie Müller, Natasha Hagadone, Yasmina Okan, Dafina Petrova, Saima Ghazal, Azeem Raza, Katrina Ellis, Margo Woller-Carter, Erich Petushek, Jinan Allan, Vincent Ybarra, Madhuri Ramasubramanian, Jinhyo Cho, Barnabás Szászi, Jessica Pope, Jonathan Huck, Sriram Govindarajan, Brandon Perelman, Brenda Bergman, Ryan McAdoo, Wesley Wehde, Kylie Key, Patrick Belling, Amad J. Chohan, Kaytlyn Green, Caitlin Tobin, Elonte' Davis, Ricardo Palma Fraga, Jahnavi Dirisina, John Wong, Angela Merritt, Florence Ettlin, Nico Müller, Anna Koehler, Niklas Keller, Amanda Gilchrist, and Joel Suss. We'd also like to thank the class on Ethical Interactions at the University of Oklahoma in the Fall of 2022 for feedback including Oliva Perrin, Will Curth, Long Nguyen, and Jordan Norris. We'd like to give a big thanks to Alantis Baldwin for copy editing, checking references, and creating the index. Financial support for publication was provided by the Dodge College of Arts and Sciences, the University Libraries, and the Office of the Vice President for Research and Partnerships, University of Oklahoma.

Finally, we'd like to thank our partners, Silke Feltz and Whitney Robison Cokely, and our families and extended families. It was a long and sometimes difficult process, but with their encouragement and support we got the time we needed.

Competing Interests. During writing this work, at least one of the authors has received the following funding:
DISES: Conservation incentives and the socio-spatial dynamics of water sustainability. National Science Foundation.
What makes climate science products useful? Exploring how stakeholders use, understand, and feel about them. SC-CASC (USGS):

Animal wellbeing and the Oklahoma/Great Plains perspective—Research in support of strategy. Kirkpatrick Foundation:
EFRI E3P: Tuning Catalyst Design to Recycle Mixed Polymer Streams. National Science Foundation.
Carbon-free H2 Production and Storage (CHEPS). Big Ideas Challenge, University of Oklahoma.
Understanding Rodeos: Education, Policy, and Attitudes Concerning Animals in Entertainment. UCLA Law School Animal Law and Small Grants Program.
Risk Communication with Partners: Your Guide to Resources and Recommendations. Funded by Medscape, USA (Medscape ID #: SF232838).
Understanding Consumer Literacy about Milk, UCLA Law School Animal Law and Policy Small Grants Program.
The Outcome Evaluation of Positive Peer Group for American Indians with Substance Use Related Offense, Portage Health Foundation, Research Excellence Fund.
Knowing What You Eat: Measuring the Effectiveness of Educational Interventions on Animal Consumption. Animal Charity Evaluators.
The Heuristics of Virtue—Integrating Virtue Ethics and the Science of Heuristics. A Science of Virtue grant from Templeton Foundation and the Arete Initiative at the University of Chicago.
Risk and Resilience of Bridges: Toward Development of Hazard-Based Assessment Framework, Research Needs, and Benefits of Accelerated Construction. University of Oklahoma Transportation Center.
Assessing Numeracy Skills of Forecasters, Partners, and Publics to Improve Tropical Cyclone Product Uncertainty Communication, IDSS, and Training. U.S. Dept. of Commerce, National Oceanic and Atmospheric Administration.
Improving web-based communications about cervical cancer screening using graphs: an interdisciplinary approach. Cancer Research, UK.
Virtual Reality Offshore Operations Training Infrastructure: Enhancing Expert Containment, Decision Making, and Risk Communications. National Academies of Science, Engineering, and Medicine.
CAREER: Numeracy and Risk Literacy. Division of Decision, Risk, and Management Sciences (DRMS) at the National Science Foundation.
Helping Doctors and Their Patients Make Decisions about Health. Ministerio de Ciencia e Innovación (Spanish Ministry of Science and Innovation).

Hands-on Abilities: Why it matters and how to improve it. Division of Engineering Education and Centers of the National Science Foundation.
Practical Wisdom as Heuristic Processes. A grant from Templeton Foundation and The Arete Initiative at the University of Chicago.
Risk Communication in Dementia. UK, Health and Social Care Board.
Baseline of Public Responsiveness to Uncertainty in Forecasts. U.S. Department of Commerce, National Oceanic and Atmospheric Administration
Influences of Affect on Philosophical Intuitions (Time-Sharing Proposal ID TESS-0090). Time-Sharing Experiments in Social Sciences funded by the Social, Behavioral, and Economic Sciences Directorate of the NSF.

Ethics Approval All the studies involving human subjects reported in this book have been approved by an IRB and participants were provided with informed consent to participate in the studies.

Contents

About the Authors

Adam Feltz serves as Professor of Psychology at The University of Oklahoma. He specializes in theoretical and applied science for ethical and informed decision making. He is regarded as one of the world's experts on the psychology and philosophy of ethical disagreement. He has published more than 60 papers on topics ranging from assessment of decision biases in surrogate decision making to the design of ethical decision support and risk communications in health, medicine, law, finance, food manufacturing, cybersecurity, natural hazards, and other domains. He is also an award-winning teacher and scholar whose research has been supported by agencies such as the National Science Foundation, the UCLA Law School, and the Templeton Foundation. He serves as a co-founding member and co-director of *RiskLiteracy.org* and is a member of the OU's Center for Applied Social Research and a member of the editorial board at *Journal of Experimental Psychology: Applied.*

Edward T. Cokely serves as Presidential Research Professor and Professor of Psychology at The University of Oklahoma. He specializes in human cognitive abilities and decision making, and has been recognized for his research on *Risk Literacy* (i.e., the ability to evaluate and understand risk), including development of the *Berlin Numeracy Test* and *Skilled Decision Theory.* Dr. Cokely has published more than 80 scientific papers and has received 22 research and teaching honors, including premier awards for "major contributions to the sciences of mind, brain, and behavior" and for "improving our understanding of the needs and processes of diverse decision makers in more than 50 countries" (FABBS, NSF). His research has been featured in Scientific American, New Scientist, Chronicle of Higher Education, BBC Futures, and the New York Times and Wall Street Journal Online, among others. He is also a dedicated educator, mentor, and collaborator, and a co-founder of *RiskLiteracy.org*, which has been promoting transparent science for informed decision making with partners from around the world, since 2012.

List of Figures

List of Tables

CHAPTER 1

Introduction

How should you live your life? Considering a wide range of possible perspectives, decision theory offers a simple prescription: Just make decisions that get you more of what you *should* want. It's a very straightforward recommendation. Nevertheless, it's hard to overstate the transformative influence of decision theory and its components, including probability theory and statistics. It seems likely that nearly every living person has felt decision theory's influence in many ways (e.g., it is an essential foundation of modern science and engineering). And with each passing day the influence of decision theory seems to be accelerating thanks to increases in knowledge, connectivity, and computing power. Yet despite its growing impact, decision theory cannot tell us what decision we should make unless we know what we should want, or more precisely *what we should value*. This limit presents serious challenges because a growing body of evidence indicates that humans have abiding philosophical biases that give rise to entrenched and fundamental disagreements. Some of these heritable biases are so resistant to change that it is unlikely we will ever come to consensus about the "truth" of many pressing moral and ethical debates—assuming one even exists. In at least some cases, philosophical biases and disagreements continue to persist even among verifiable experts who have devoted their lives to understanding and clarifying all relevant issues and facts. The heritability of our philosophical biases may also help explain why some philosophical debates have persisted generation after generation. But make no mistake, these empirical claims do not reflect mere

A. Feltz, E. T. Cokely, *Diversity and Disagreement*,
https://doi.org/10.1007/978-3-031-61935-9_1

armchair conjecture. Instead, they follow from intensive scientific inquiry that has unfolded over the last two decades, which has revealed consistent and converging evidence on the fragmentation and potential immutability of some of our most fundamental philosophical beliefs. As is detailed throughout this book, primary evidence can be found in nearly 100 scientific studies involving thousands of diverse participants from many cultures and countries around the world. This research confirms what some have argued for centuries: Even for the most informed and reasonable people, it may be impossible to ever agree about some of humanity's most defining values and moral issues.

The scientific research documenting the robust nature of some philosophical disagreements provides a foundation for a formal axiomatized normative argument that has established new, strong bounds on justifiable philosophical practice and inference—i.e., the Philosophical Personality Argument. The core of this book revolves around the Philosophical Personality Argument, which is:

1. "Philosophically relevant intuitions are used as some evidence for the truth of some philosophical claims.
2. Some differences in philosophically relevant intuitions used as evidence for the truth of some philosophical claims are systematically related to some differences in personality.
3. If philosophically relevant intuitions are used as some evidence for the truth of some philosophical claims and those intuitions are systematically related to some differences in personality, then one's endorsement of some philosophical claims is at least partially a function of one's personality.
4. Therefore, one's endorsement of some philosophical claims is at least partially a function of one's personality." (Feltz & Cokely, 2012a)

We will spend the bulk of this book providing evidence for the truth of the premises (1–4) and defending them from various objections. We then discuss some of the potential implications of the Philosophical Personality Argument. Among its many implications, the Philosophical Personality Argument casts light on how and why those who engage in some efforts to determine the non-conceptual, non-linguistic truth of various fundamental philosophical and moral issues are likely to generate seemingly accurate, yet irreducibly biased and diverse conclusions. This formal approach also implies that we may be unlikely to ever have access to uncontroversial, unequivocal answers to many fundamental philosophical and ethical questions that are believed to underwrite so much of what we humans value most dearly. Moreover, the Philosophical Personality Argument indicates that those who engage in other influential projects—such as normative projects or conceptual analyses that use intuitions as essential, irreplaceable elements—also bear considerable risk for undetected bias and error in their scholarship. Beyond the obvious limits and implications that follow for *philosophers, ethicists, scholars, and policy makers*, we explore other implications for high-stakes decision making interventions and human welfare more broadly. Accordingly, in the last chapter our primary focus is an examination and exploration of key implications, culminating in a new normative theory and scientific framework for the science of informed decision making—i.e., *Ethical Interaction Theory.*

A Day in the Life

To illustrate how we came to such strong conclusions, it seems useful to look at some typical activities that philosophers and psychologists engage in. Accordingly, we'll invite you to do some philosophy and to explore your own psychology by making some judgments and decisions. We'll also put some skin in the game and make research-based predictions about your reactions to some paradigmatic philosophical scenarios based on your self-reported personality traits.

For a philosopher, a typical day at the office involves reading a lot. Much of this reading involves thought experiments on difficult problems. Many of these problems are about issues that have deep and meaningful implications for most people. For example, consider the following scenario:

> Imagine a universe (Universe A) in which everything that happens is completely caused by whatever happened before it including human decision making. For example, one day Mary decided to have French Fries at lunch. Since a person's decision in this universe is completely caused by what happened before it, given the things that happened in Universe A before her decisions, it *had to happen* that Mary would decide to have French Fries. She could not have decided to have something different.
>
> Imagine in Universe A a man named Bill has become attracted to his secretary, and he decides that the only way to be with her is to kill his wife and 3 children. He knows that it is impossible to escape from his house in the event of a fire. Before he leaves on a business trip, he sets up a device in his basement that burns down the house and kills his family.

Take a few minutes to re-read the scenario and think about some of its key features. Once you've done that, please answer this question:

"Is Bill morally responsible for killing his family?"

Please go ahead and write down your answer to that question—we will return to this question later.

You have just engaged in one kind of typical philosophical activity. Of course, a lot of other things happen in a typical day as a philosopher, but thinking about cases illustrating philosophical ideas is certainly one of them. You've thought deeply about a case and have made a judgment about that case. That judgment, in turn, probably reflects some of the deep values that you have. To us, this exercise illustrates a wonderfully democratic aspect of philosophy. Almost all people can make judgments about the case you just read. So, in that sense, we can all take part in philosophical activity. Moreover, philosophers often take what people think very seriously when they construct their philosophical theories. After all, philosophers often want to think and theorize about things that are of central importance to human existence. Since you are an existing human, your thoughts about these issues matter too.

But there's a challenge associated with this democratic spirit of philosophy. Suppose somebody disagreed with you about the answer to the statement "Bill is morally responsible for killing his wife." How would you go about trying to convince that person that they are wrong? You would probably reference key elements of the scenario to justify your view, and you might then argue that those elements are the reasons why a person in that position would be as free as you judged them to be. But what would you do if you still could not convince the other person that they are wrong? What if the person with whom you disagree highlights other aspects of the scenario? Is that disagreement simply unresolvable? Does that imply that at least one of you is making a mistake? And how could you tell if one of you was making a mistake? Those are difficult, key questions that we'll return to again and again in this book.

Shifting gears a little bit, what comes to mind when you think of a psychologist? You might be imagining somebody who invites people to sit on a couch to discuss distressing feelings or resolve personal problems. Alternatively, you may have thought of a person in a lab coat testing participants and assessing their behavior. For our current purposes, it's the lab coat type psychologist that we would like you to focus on. A typical day for these kinds of *research psychologists* (like us) often involves trying to more precisely understand human psychology and behavior via empirical investigation and experimentation. For example, we sometimes create and test technologies designed to measure or improve cognitive abilities (e.g., learning, reasoning, decision making). Other times, we examine how risk communications about recycled water or natural hazards can influence our emotions, attitudes, and choices. And, of course, we also investigate the reasons people disagree about philosophically relevant judgments. But again, unlike most philosophers, psychologists primarily conduct *empirical studies and experiments* to investigate the reasons why people make the judgments they do. Psychologists usually also develop and validate instruments (e.g., tests, surveys, training systems) and create mathematical models of behavior, which allow them to make predictions about how and why people will behave the way they do in the future.

Philosophical Personality

Personality is one influential factor that psychologists have used to help explain and predict how people feel and what they think or do (Revelle & Scherer, 2009). You probably already have a sense for this. You know people who are outgoing and social. You can make reasonably accurate predictions about how these people will behave in different situations. You may even deliberately use ideas about people's personality to help inform your predictions. Psychologists do the same, although they typically use systems that are more precise and scientifically grounded, based on taxonomies of personality traits. Some of the more well-known assessments developed by psychologists include the Big Five personality traits, the HEXACO traits, Myers-Brigs personality traits, and the Minnesota Multiphasic Personality Inventory (MMPI), to name just a few. Each of these approaches provides unique insights that can help predict patterns of feelings, thoughts, and behaviors. For example, the MMPI is often used to help clinicians identify and treat psychopathologies and related mental health challenges. Given our current aims, in this book we will mainly focus on what we take to be the most influential approach to adult personality traits, namely the Big Five personality model (sometimes referred to as the OCEAN traits).

The Big Five model of personal includes five "global" personality traits: Extraversion, openness to experience, emotional stability, agreeableness, and conscientiousness. The gist of each global trait is pretty much what you would expect from the labels. Extraverts enjoy social interaction, agreeable people tend to avoid conflicts, and so on. We will discuss some of the Big Five traits in greater depth in future chapters, but for readers who want to begin with a more comprehensive overview we recommend the review by John and Srivastava (1999). One thing that will be apparent in any high-quality review is that there is substantial empirical evidence showing that the Big Five personal traits are heritable (i.e., related to differences in people's genes), which partly explains why they robustly predict patterns of feelings, behaviors, and thoughts. For example, people who are higher in the heritable trait of extraversion are more likely to take on and enjoy leadership roles (Judge, Bono, Ilies, & Gerhardt, 2002). Of course, this doesn't mean that being highly extraverted will guarantee that a person will volunteer for, or enjoy, leadership roles. But all else equal, the evidence suggests that it's a good bet they will.

The Big Five personality traits have also been found to be exhibited by diverse people from many unique cultures all around the world. They have been found to be relatively stable once a person hits adulthood. They also encompass a wide swath of more specific tendencies, called facets, which can provide more nuanced insights into people's feelings and judgments. For example, extraversion is a global trait that typically encompasses individual differences in the facet of warmth—i.e., the tendency to be close and affectionate with others. This suggests that although people who are more extraverted also tend to be warmer than introverts, some extraverts are much warmer than other extraverts. As such, when it comes to philosophical and moral questions, sometimes the specific facet of personality (i.e., warmth) may be a stronger predictor of someone's judgment than their more general global trait (i.e., extraversion). Given these and other findings, one of the central claims we will present in this book is that some of these heritable Big Five personality traits, and the specific facets thereof, help to explain (and predict) at least some noteworthy philosophically relevant judgments.

Know Yourself (and Others)

Personality tests have been shown to predict many things about people such as longevity, career outcomes, and success in marriage. If you are interested in your personality traits, you can find tests online to determine how you score on the dimensions of the Big Five. Or, you can take this quick personality test created and validated by Gosling, Rentfrow, and Swann (2003). This short ten-item test is a measure of each of the Big Five global personality traits. This measure has remarkably robust predictive power even when compared to much longer instruments that measure fine-grained facets of each personality type. To give it a try, simply rate yourself on each of the following pairs of adjectives, using a scale numbered from 1 through 7. Use 1 to indicate strong disagreement or on the opposite end of the scale use 7 to indicate strong agreement (e.g., an answer of 5 would indicate mild agreement with that pair of adjectives whereas 3 would indicate mild disagreement, and so on).

1. Extraverted, enthusiastic ________
2. Critical, quarrelsome ________
3. Dependable, self-disciplined ________
4. Anxious, easily upset ________
5. Open to new experience, complex ________
6. Reserved, quiet ________
7. Sympathetic, warm ________
8. Disorganized, careless ________
9. Calm, emotionally stable ________
10. Conventional, uncreative ________

Once you rate your agreement with these ten pairs of adjectives, you can use the following instructions to score your personality on each of the five personality factors. But please note, there's one tricky part to calculating your score on this brief Big Five inventory. You will need to reverse score questions 2, 4, 6, 8, and 10. The basic idea behind reverse scoring is simple—take the mirror image of your score.[1] So, if you rated yourself a 1 for items 2, 4, 6, 8, or 10, then your reverse score for those items is a 7, if you rated yourself a 2 your score is 6, if 3 your score is 5, 4 says the same, but 5 then becomes 3, a score of 6 becomes 2, and a score of 7 is transformed into a 1.

1. Extraversion (sum of 1 and reverse score of 6) ______
2. Agreeableness (sum of reverse score of 2 and 7) ______
3. Conscientiousness (sum of 3 and reverse score of 8) ______
4. Emotional Stability (sum of reverse score of 4 and 9) ______
5. Openness to Experience (sum of 5 and reverse score of 10) ______

[1] At the time of publishing, the Gosling lab has a spreadsheet you can use that will calculate the scores for you here: http://gosling.psy.utexas.edu/wp-content/uploads/2014/09/excelscoreTIPI.xls.

Here is where we are going to take a small risk in this book. We are going to try to predict how you will respond to a number of different philosophical scenarios that involve philosophically relevant concepts like freedom, morality, and intentional action using your scores to these global personality traits. We're confident we won't always be right. But odds are that on average we'll do fairly well for most people, so let's give it a shot. You've already responded to a paradigmatic scenario probing your thoughts about some aspects of freedom and moral responsibility. The next step is to consider and make some judgments about the next two scenarios as well:

> Suppose the vice-president of a company went to the chairman of the board and said, "We are thinking of starting a new program. It will help us increase profits for this year's balance sheet, but in ten years it will start to harm the environment." The chairman answered, "I don't care at all about harming the environment. I just want to make as much profit for this year's balance sheet as I can. Let's start the new program." They started the new program. Sure enough, the environment started to be harmed. (Knobe, 2003a)

Think about that scenario for a few minutes. Then, write down your answer to the following question: Did the chairman intentionally harm the environment? Now, imagine a slightly different scenario (pay attention—there is a subtle difference: To help, we've put the difference in bold):

> Suppose the vice-president of a company went to the chairman of the board and said, "We are thinking of starting a new program. It will help us increase profits for this year's balance sheet, but in ten years it will start to **help** the environment." The chairman answered, "I don't care at all about **helping** the environment. I just want to make as much profit for this year's balance sheet as I can. Let's start the new program." They started the new program. Sure enough, the environment started to be **helped**. (Knobe, 2003a)

Answer the following question: Did the chairman intentionally help the environment? Given our research, if you scored high on extraversion you are likely to think that the chairman intentionally harmed the environment yet you are also quite likely to say that the chairman *did not* intentionally help the environment. If you scored lower on extraversion, you'll likely think neither was intentional. So, who is right? Some people think they know the answer… but we're not so sure. In any event, we think differences in intuitions about intentionality revealed using cases like these may tell us something important about the origins and tractability of related higher-stakes disagreements (e.g., court cases about liabilities, criminal penalties, and fair compensation; expectations for praise and blame in personal and romantic relationships).

We'll give you one last scenario as a taste for what is to come in the chapters that follow.

> Imagine that John and Fred are members of different cultures, and they are in an argument. John says, "It's okay to hit people just because you feel like it," and Fred says, "No, it is not okay to hit people just because you feel like it." John then says, "Look you are wrong. Everyone I know agrees that it's okay to do that." Fred responds, "Oh no, you are the one who is mistaken. Everyone I know agrees that it's not okay to do that."
>
> Suppose somebody asks you who is right in the debate? Is Fred right that it is not okay to hit people just because you feel like it? Is John right that it is okay to hit people just because you feel like it? Or is there no fact of the matter about claims like hitting others just because you feel like it? (Nichols, 2004a)

Just like the previous scenarios, please take a few minutes to carefully consider the key elements of the scenario. Then, write down your answer to the question. In this case, based on previous research, we predict that if you are high in openness to experience you are more likely to think that there is no fact of the matter about hitting others just because you feel like it. However, if you are lower on openness to experience, you're more likely to have other views. So, how did we do? Odds are we probably did well "overall, and on average," as the saying goes. Regardless of our prediction, you might be wondering whether you gave the "right" answer. Again we really can't tell you. Even though we have our own personal opinions, we just don't know. That said, we feel confident that some esteemed philosophers and psychologists probably would argue that they indeed do know the right answer… but ironically if we could survey enough of these people who "know" the right answer we'd likely find that there is strong disagreement about what the "obvious" right answer "must be." To us, that disagreement is a very interesting finding, particularly because we can use people's heritable personality traits to predict what they will think. As such, much of this book is devoted to explaining why we make the predictions that we do, and what we think these findings mean for philosophy, psychology, and people's lives more generally. We hope it is obvious that we take your responses to these kinds of scenarios seriously. We also take the relation of your responses to your personality seriously. And, if you decide you want to get more substantial feedback on your own philosophical personality profile, we invite you to visit our website at PhilosophicalCharacter.org. It has options to provide feedback about how you compare to others, and what implications your specific philosophical personality profile may have, and sometimes there are opportunities to volunteer to participate in studies in the future. Just please remember, we're much better at research than webpage design, so assuming you don't expect too much, you'll probably be pleasantly surprised.

A Road Less Traveled

What you just did—answer a personality inventory and answer some philosophically relevant questions—is exactly how we started down the road to discovering the relations between philosophically relevant judgments and personality. Technically, we can't quite remember (or agree about) how exactly we decided to embark on our research collaboration, which started almost 20 years ago at Florida State University. It seems likely that at least one of us was just a little more extraverted than normal while we were waiting for a research meeting to begin. One version of what happened next is that we struck up a discussion about statistics that resulted in a question about the role of individual differences in philosophical judgments. From there our work together began in earnest with an investigation of free will judgments. We thought that there might be some relation between some free will judgments and personality (i.e., extraversion) but we couldn't find any direct test in the literature. When we discovered the relation, we were encouraged. We then started making and testing predictions about a host of philosophically relevant judgments. Although our work involved elements of psychology and philosophy, our research built primarily on the scholarly work that has come to be known as "experimental philosophy." Experimental philosophy typically involves using some methods from psychology (and allied behavioral and social sciences) to help address some philosophically relevant questions. At the time we first started working on these issues, free will, intentional action, and moral judgments were among the dominant themes explored in experimental philosophy. Consequently, our research mainly focused on those main three domains.[2]

The relation between personality and philosophically relevant intuitions has many implications for theory and can also have some notable practical implications. The first theoretical implication we will discuss in this book is that the observed relationship with personality suggests that for some notable philosophical issues there probably is not any single uncontroversial "folk" view (e.g., people who feel differently about important issues can and probably should reasonably disagree on some of the philosophical issues). Nevertheless, many philosophers talk as if there is or should be a single folk view when making arguments about the "truth" of some philosophical topics (e.g., "most people would agree that…"). Taking that kind of "one size fits all" approach to the perspectives of diverse people around the world is naïve, it is contrary to a large body of scientific evidence, and it is obviously inconsistent with the characteristics of many modern political, social, and moral disagreements. From our perspective, these findings suggest that it may not be as important to try and find the "right" view, but rather to understand the reasons for philosophical disagreements, and to start to map the implications of people's reasonable yet diverse views.

Second, the evidence we present in this book might at least partially explain why some philosophical debates have never been resolved. That is, some enduring debates may never go away simply because philosophers have different personalities, and as such they are likely to have heritable philosophical biases that shape how they feel about fundamental moral and ethical issues. Once again, because personality traits are heritable and related to core philosophical intuitions, debates that involve those intuitions seem unlikely to result in agreement about uncontroversial "correct" views about "the" truth of some philosophical issues. Moreover, this finding indicates that some philosophical methods are not nearly as reliable as previous thought, and thus many traditional philosophical projects that use intuitions as evidence should become substantially more empirically oriented, and should also become much less focused on finding "the" Neo-Platonic truth (if such a thing

[2] Of note, we also looked at personality relations to some effects in epistemology given that one canonical paper in experimental philosophy is the Weinberg, Nichols, and Stich (2001) paper. We found no reliable relations to personality, but this may be perhaps because the effects reported in that paper have been hard to replicate. We give some extensive discussion along these lines in the next few chapters.

even exists). While this may seem like a somewhat pessimistic view, one that could also complicate the work of some philosophers in unwelcome ways, we remain optimistic for many reasons. First and foremost, we're after the truth, even if it's inconvenient. Secondly, by our lights this finding appears liberating and potentially empowering. After all, if we're all biased and thus fundamentally disagree about some important issues, no matter how much we argue we're just not going to agree... but as many have noted disagreement doesn't mean we can't find productive ways to respectfully consider other people's perspectives. Accordingly, we end the book by arguing that given the increasing power of behavioral science to control our behaviors, there is a need to reconsider how we design and evaluate the technology and policies that increasingly influence our choices (e.g., the nudging versus boosting debate in behavioral economics, politics, and business).

Guide to Reading the Book

In Chap. 2, we begin to document the empirical evidence concerning the relation between personality and philosophically relevant intuitions. In many ways, Chap. 2 serves as the primary example of personality predicting intuitions about philosophically relevant judgments. In Chap. 2, we document the evidence that global personality traits predict intuitions about compatibilism, fatalism, and manipulation.

In Chap. 3, we detail personality's relation to intentional action intuitions. These intuitions include judgments about side effects of actions that are illustrated by the two chairmen cases that you have already responded to above. We go on to demonstrate that this effect persists even after controlling people's concepts of intentional action and for different kinds of materials and testing environments.

Chapter 4 provides evidence that personality predicts some ethically relevant intuitions. These intuitions range from intuitions about meta-ethical positions, first-order ethical positions, and applied ethical positions. One such example is the one you responded to above—whether there are facts of the matter about some ethical claims. But personality also predicts people who are likely to attribute virtues to others, and also predicts some applied ethical judgments about punishment, moral wrongness, and desert.

Chapter 5 provides an extended argument concerning one of the important objections to the kinds of claims we are making—the Expertise Defense. The existing literature on expertise suggests that in many domains and for many judgments, expertise makes a real, qualitative difference. Experts in those domains are simply better than non-experts in important respects. We detail some of the arguments, theories, and evidence about why experts often make better judgments than non-experts. We go on to argue that philosophy is not likely to be one of those domains, at least for many of the judgments that are of interest. We also provide an empirical test of the expertise defense and find that personality predicts intuitions about freedom and moral responsibility even among verified experts. These results suggest that for many philosophical domains and judgments, the expertise defense fails.

In Chap. 6, we present the Philosophical Personality Argument that you encountered at the beginning of this chapter. In short, the argument presents an axiomatized, deductive, normative argument based on the diversity of philosophical intuitions associated with personality. Given that personality is irrelevant to the truth of the content of intuitions, we argue that some philosophical practices that attempt to establish the mind-independent, non-linguistic, non-conceptual truth using those intuitions run the risk of not being able to succeed. We take this chapter to be the philosophical core of the book.

Chapter 7 takes the key insights from the Philosophical Personality Argument and draws practical implications. Even though philosophical values are importantly diverse, there is large consensus among people and cross-culturally that there are some core values. Autonomy and beneficence are two core values that have received relatively large consensus. Striking a balance between these two values has been challenging and the subject of much debate. One important and popular approach to negotiating this balance is *Libertarian Paternalism*. Libertarian Paternalism attempts to help people engage in better behaviors (thereby promoting beneficence) while at the same time protecting freedom of choice (thereby respecting autonomy). However, we argue that there are significant ethical costs associated with Libertarian Paternalism that are often neglected, especially when considered in light of systematic diversity of philosophical values. To help address these concerns, we present a new complementary theory, *Ethical Interaction Theory,* and argue that it is likely both necessary and useful for studying human interactions and designing choice architecture (e.g., user interfaces), that respects and promotes autonomy and diversity. This approach attempts to determine the unique set of capabilities and values that are involved in (good) decisions in order to help people integrate those things into their own decision making processes, in accord with their own values, beliefs, needs, and responsibilities. To support practitioners (e.g., designers, psychologists, engineers), we also provide a theoretically grounded, practical checklist to compare the relative merits of different kinds of choice architecture (e.g., Libertarian Paternalistic versus Informed and Independent).

There are a few approaches you may want to consider for reading this book depending on your interests. For those who are interested in the theoretical and philosophical implications of the relation of personality with philosophical intuitions, you can safely read Chap. 6 alone (or at least start there). For those of you who are only interested in policy debates and ethical interaction design, you can skip directly to the last chapter. For those of you who are interested in evidence on the empirical findings about personality predicting philosophical intuitions, you can read Chaps. 2–5.

The primary goal of this book is to help provide a theoretical framework for understanding variation in fundamental philosophical intuitions, and to carefully consider how that variation may help promote the development of more ethical interactions. Along the way, we hope you will come to agree that it is no longer scientifically responsible or ethically defensible to treat all people as if they could or should normally have the same philosophical intuitions or values. Additionally, we hope you will see why we feel so strongly that people should not normally try to influence other people's decisions without also giving careful consideration to diversity and disagreement in philosophical intuitions.

CHAPTER 2

Freedom and Responsibility

"Suppose scientists announced today that scientists discovered that all human behavior is entirely caused by previous events. This would mean that whenever a person acts, that action is completely caused by events that occurred earlier in a person's life, and those events are also completely caused by even earlier events eventually going back to events that occurred before a person was born. This also implies that events that occur before a person is born are part of a sequence that will definitely cause all the actions and decisions that person makes. Now imagine that John decides to cheat on his taxes and does it. If the scientists are right, then John's decision to cheat on his taxes is completely caused by a series of events that started before he was born. So, the question is: Did John decide to cheat on his taxes freely? Is John morally responsible for cheating on his taxes?"

Imagine you found out you have no free will—you're something like a robot or a hologram, and all your actions are precisely scripted and dictated by your programming. Would this knowledge change how you live your life or what you value? Should it change the way you treat other people or the way you feel about your successes or failures? Some people spend virtually their entire professional careers thinking about these and related issues. Many theorists argue that beliefs about free will are essentially related to our conceptions of ourselves and our relationships with others. These conceptions and beliefs in free will run deep and are thought by some to underwrite our notions of justice, punishment, desert, and self-worth (Kane, 1996). There is gathering experimental evidence that perhaps these views are at least in part correct where decreasing belief in

A. Feltz, E. T. Cokely, *Diversity and Disagreement*,
https://doi.org/10.1007/978-3-031-61935-9_2

free will can be related to increases in cheating behaviors (Vohs & Schooler, 2008) and worse job performance (Baumeister, Masicampo, & DeWall, 2009; Baumeister, Sparks, Stillman, & Vohs, 2008; Stillman, Baumeister, & Mele, 2011).[1] Some think that these connections run so deep that if we are in fact not free or morally responsible, we should allow people to continue to have a false belief about their freedom and moral responsibility (Smilansky, 2000). If these theorists are right, then our understanding of what is required for freedom and responsibility forms a cornerstone of our understanding of ourselves and our relation to the world.

In this chapter, we provide an overview of some of the classic findings in the experimental philosophy of free will. These classic findings suggest that people's intuitions about freedom and moral responsibility are at least sometimes related to a variety of factors including the affective content of the scenarios and how determinism is described. However, we focus primarily on results from our research program indicating that in a wide variety of instances, people's free will and moral responsibility intuitions are associated with general and heritable personality traits. In later chapters, we will argue that the relation between personality and free will and moral responsibility challenges some long-standing assumptions held both by traditional and experimental philosophers.

The Experimental Philosophy of Free Will

It is common for philosophers to take intuitions that are pervasive among non-professional philosophers seriously. The intuitions of philosophical non-experts are sometimes referred to as "folk intuitions." There is some debate about what intuitions actually are, but they are generally thought to be immediate reactions or judgments that one has about concrete situations or scenarios (see, for more details, A. Feltz and Bishop (2010) and Chap. 6). Some have argued that positions supported by folk intuitions have "squatter's rights" (Dennett, 1984; Nahmias, Morris, Nadelhoffer, & Turner, 2006b). That is, those who endorse philosophical views that are inconsistent with folk intuitions shoulder an additional argumentative burden to explain why those intuitions are mistaken. Those who have views consistent with folk intuitions do not shoulder an additional argumentative burden. Of course, a theory *need not* respect folk intuitions about freedom and moral responsibility. For example, those who have views about free will and moral responsibility that are inconsistent with

[1] Some of these effects have been difficult to replicate. See, for example, Nadelhoffer, Shepard, Crone, Everett, Earp, & Levy (2020b).

folk intuitions could offer a revisionary account of free will and moral responsibility (Vargas, 2005). But even in revisionary views, leading authorities agree that some account of folk intuitions is desirable. Precedent entails that if folk intuitions about freedom and moral responsibility play this role (e.g., determining squatter's rights), then they can have some substantial role to play in philosophical theorizing about free will and moral responsibility. The same way a representative in Congress would ideally want to proxy the interests and intent of their constituents, so too should the ethicist, philosopher, and legal scholar somehow represent the folk in their analysis.

Often philosophers, bioethicists, legal scholars, and others make an explicit and direct appeal to folk intuitions about freedom and moral responsibility to support their views (e.g., Beauchamp and Childress (2009); Dennett (1984); Kane (1996); Pink (2004); Sommers (2010); Strawson (1986)). Most adults make free will and moral responsibility attributions and judgments routinely and without much difficulty. This everyday practice is of primary interest of scholars and philosophers of free will—it is the phenomenon that many philosophers are interested in understanding, analyzing, and theorizing about. Theories of free will and moral responsibility that are not constrained by everyday practices run the risk of being "philosophical fictions," (Mele, 2001)—views that may be internally consistent but do not refer to anything in the real world or of value to people. Some scholars who think that philosophical theorizing about freedom and moral responsibility should be constrained by folk intuitions assume that they have a fairly good understanding of everyday attitudes about freedom and moral responsibility. For example, talking in general terms about intuitions, Jackson writes "it is also true that [professional philosophers] often know that our own case is typical and so can generalize from it to others" (1998, p. 37).

As noted in the introduction, a growing body of research indicates that making reference to folk intuitions to support philosophical claims about free will is more tenuous and complicated than might have been thought. One obvious sign of the difficulty is that sometimes theorists disagree about what everyday intuitions about freedom and moral responsibility are. For example, some think that everyday intuitions are compatibilists (i.e., free will is compatible with the truth of determinism) whereas others think that everyday intuitions are incompatibilist (i.e., free will is not compatible with the truth of determinism) (Dennett, 1984; Ekstrom, 2002; Kane, 1996; Lycan, 2004; Pink, 2004; Strawson, 1986; Wolf, 1990). On the face of it, not all these views are accurate descriptions of folk intuitions—folk intuitions cannot primarily support compatibilism and incompatibilism. Determining which views best describe what intuitions people in fact have about freedom and responsibility is efficiently done using empirical methods of the behavioral sciences—an approach that has been dubbed "experimental philosophy."

The scenario at the start of this chapter is a research instrument designed to represent key elements of determinism. Determinism is the thesis that "at any instant exactly one future is compatible with the state of the universe at that instant and the laws of nature" (Mele, 2006, p. 3). The question of whether John is free and morally responsible is known as *the compatibility question*—a fundamental question that has taken center stage in the contemporary free will debate (Kane, 1996; Sommers, 2010). Compatibilists think that the answer to the compatibility question is "yes" because they hold that free will and moral responsibility are compatible with determinism. Incompatibilists think the answer is "no," John is not free or morally responsible. Theorists predictably disagree. On the one hand, some think that John is not morally responsible because his decision to cheat on his taxes was completely caused by a series of events that extends back in time to before he even was born. And, the thinking goes, John cannot be morally responsible for or freely do things that happened before he even existed. So, he cannot be responsible for anything that is completely the result of those events (Strawson, 1994; Van Inwagen, 1983). Given the past and the laws of nature, there is nothing John could have done not to cheat on his taxes. On the other hand, John is a complex individual who makes decisions based on *his* particular set of desires and beliefs. Even if those desires and beliefs were completely determined by other factors, he was not coerced or forced into cheating on his taxes—his cheating on his taxes is an expression of who he is and what he values (Frankfurt, 1971; Watson, 1975). So, for these reasons, one might think that John acted freely and is morally responsible for cheating on his taxes.

Some of the early empirical work suggested that everyday intuitions supported compatibilism. But asking for people's intuitions about determinism's relation to freedom and moral responsibility is no easy feat. Determinism is a technical term that not many people completely understand the way that philosophers do. Because of the challenges associated with describing technical philosophical concepts, researchers often use scenarios that capture central elements of those concepts. These scenarios can convey some of the key elements of determinism to non-experts without having to use technical jargon. To illustrate, Nahmias, Morris, Nadelhoffer, and Turner (2005) presented participants with several scenarios describing a person acting in a deterministic universe. Almost no one thinks that determinism actually describes the causal processes of our world (see Nichols and Knobe's study below). So, Nahmias and colleagues described a hypothetical supercomputer that knows all the laws of nature and has a complete description of the universe at a given time. From these facts, the supercomputer infallibly deduces everything that will happen in the future. In a determined world, nothing is in principle unpredictable, so the supercomputer is meant to illustrate a central element of determinism. They then asked participants if the person freely performs and is morally responsible for an action in that world. One of their studies involved a person named John who robs a bank.[2] The supercomputer predicts that John will rob a bank at a precise date and time, and John robs the bank at that exact moment. In this case, the majority of participants (more than 75%) thought that John freely robbed the bank and was morally responsible for robbing the bank. Since most people judged that John was free and morally responsible for robbing the bank in a determined world, the results suggest that most people at least sometimes have compatibilist intuitions.

However, data have since emerged indicating folk intuitions about free will and moral responsibility can be responsive to environmental or situational factors. For example, Shaun Nichols and Joshua Knobe (2007) provide evidence that folk intuitions are influenced by whether people make judgments about some abstract individual or about a person who is described in some detail. They also found that the emotional content that a scenario has could also influence folk intuitions about freedom and moral responsibility. To assess people's intuitions about abstract questions concerning compatibilism, Nichols and Knobe (2007) provided participants with the following descriptions. Universe A is meant to describe some key elements of determinism. Universe B is meant to describe a universe where determinism is not true:

> *Universe A*: Imagine a universe (Universe A) in which everything that happens is completely caused by whatever happened before it. This is true from

[2] Nahmias et al. (2005) also presented actions that were morally good or morally neutral and used different descriptions of determinism (e.g., genes and upbringing caused all of a person's actions). The results did not dramatically vary in these other cases.

> the very beginning of the universe, so what happened in the beginning of the universe caused what happened next, and so on right up until the present. For example, one day John decided to have French Fries at lunch. Like everything else, this decision was completely caused by what happened before it. So, if everything in this universe was exactly the same up until John made his decision, then it *had to happen* that John would decide to have French Fries (Nichols & Knobe, 2007).
>
> *Universe B*: Now, imagine a universe (Universe B) in which *almost* everything that happens is completely caused by whatever happened before it. The one exception is human decision making. For example, one day Mary decided to have French Fries at lunch. Since a person's decision in this universe is not completely caused by what happened before it, even if everything in the universe was exactly the same up until Mary made her decision, it *did not have to happen* that Mary would decide to have French Fries. She could have decided to have something different.

After reading these descriptions, participants were asked which of the universes is most like ours. Not surprisingly, the vast majority (over 90%) thought Universe B (the indeterministic universe) was most like ours. When asked the following question "In Universe A, is it possible for a person to be fully morally responsible for their actions?" the vast majority (86%) said "no" (Nichols & Knobe, 2007, p. 670). However, responses changed dramatically when participants were presented with the following "concrete" paragraph in addition to the two above:

> In Universe A, a man named Bill has become attracted to his secretary, and he decides that the only way to be with her is to kill his wife and 3 children. He knows that it is impossible to escape from his house in the event of a fire. Before he leaves on a business trip, he sets up a device in his basement that burns down the house and kills his family.

Seventy-two percent of participants thought that Bill was morally responsible for killing his wife in Universe A even though his action was determined. The results suggest that people can have very different intuitions depending on whether the questions are asked in abstract, general terms about actions or about concrete, specific terms about a particular person's action.

What could explain the abstract/concrete difference? Nichols and Knobe (2007) posit that emotional reactions could explain the different intuitions about the cases. Emotions often influence people's reactions and judgments. For example, if you are feeling angry, you are less likely to adequately acknowledge legitimate excuses for other people's behavior. Imagine you had a bad day at work and then you come home and see that your dog has eaten the leftover pizza you left on the counter. Because you are already angry, you are less likely to think that you may have done something wrong to enable your dog to eat the pizza (e.g., leaving it out on the counter). You'll be more likely to blame and be angry at your dog because of your antecedent bad mood than if you were not in a bad mood. In this case, emotions get in the way of how you would assess a situation if you weren't feeling emotional. Nichols and Knobe suggest something similar may be happening in the abstract and concrete cases. Affect, or one's experience and expression of emotions, may get in the way of one's judgments about freedom and moral responsibility—one's affective state results in a performance error. According to the affective performance error account, people normally have an incompatibilist theory of freedom and moral responsibility as illustrated in the abstract scenario. However, in cases with high affective content, one's emotional reactions get in the way of one's incompatibilist theory resulting in more compatibilist friendly intuitions as illustrated in the concrete scenarios.

To test the affective performance error model, participants received the following concretely described scenarios that varied the affect content of the action in addition to the description of Universe A above.

> *High Affect:* As he has done many times in the past, Bill stalks and rapes a stranger. Is it possible that Bill is fully morally responsible for raping the stranger?
>
> *Low Affect:* As he has done many times in the past, Mark arranges to cheat on his taxes. Is it possible that Mark is fully morally responsible for cheating on his taxes? (Nichols & Knobe, 2007)

When asked if Bill could be fully morally responsible for his actions, 64% said yes. However, only 23% said that Mark could be fully morally responsible for his actions. Hence, it appears that the affective component of the action influences people's intuitions about moral responsibility's relation to determinism.

The affective performance error model seems to explain the abstract/concrete difference. Abstract cases generally generate less affect than concrete cases. In some concrete cases, the action is described in enough detail such that more affect is generated.[3] Hence, depending on the affective content of the action being evaluated, people can sometimes express compatibilist and sometimes express incompatibilist intuitions. Some recent studies have cast doubt on the affective performance error model, however (Cova, Bertoux, Bourgeois-Gironde, & Dubois, 2012; Vargas, 2006). A meta-analysis of the affective performance error model suggests that while the effect of affect is real (~1% of total variance), it is not large enough to explain the large differences between the abstract and concrete cases (A. Feltz & Cova, 2014).

Nahmias, Coates, and Kvaran (2006a) also provide evidence that people's free will intuitions are sometimes sensitive to contextual factors and offer an alternative explanation for the concrete/abstract difference. In particular, sometimes people's free will intuitions are responsive to the nature of the description of determinism. To demonstrate this, Nahmias, Coates, et al. (2006a) created vignettes where determinism was described in "psychologically reductionistic" terms or in "psychologically non-reductionistic" terms. The reductionist scenarios (in italics below) describe mental processes in terms of brain states and processes whereas the non-reductionist scenarios (in brackets below) describe mental process in more folk-psychological terms like thoughts and desires:

> Most respected *neuroscientists* [psychologists] are convinced that eventually we will figure out exactly how all of our decisions and actions are entirely caused. For instance, they think that whenever we are trying to decide what to do, the decision we end up making is completely caused by the specific *chemical reactions and neural processes* [thoughts, desires, and plans] occurring in our brains. The *neuroscientists* [psychologists] are also convinced that these *chemical reactions and neural processes* [thoughts, desires, and plans] are completely caused by our current situation and the earlier events in our lives, and that these earlier events were also completely caused by even earlier events, eventually going all the way back to events that occurred before we were born.
>
> So, if these *neuroscientists* [psychologists] are right, then once specific earlier events have occurred in a person's life, these events will definitely cause specific later events to occur. For instance, once specific *chemical reactions and neural processes* [thoughts, desires, and plans] occur in the person's

[3] Of note, the affective content of the action has not been directly measured in these studies.

brain [mind], they will definitely cause the person to make the specific decision he or she makes. (Nahmias, Coates, et al., 2006a, p. 224)

The results from these scenarios were impressive. Forty percent of those in the psychologically reductionist scenario judged the person to be free and morally responsible. Eight-five percent of those in the non-reductionistic scenario judged that the person was free and morally responsible. Nahmias, Coates, et al. (2006a, p. 229) also found the same basic pattern of results using different but conceptually similar scenarios (see also Roskies and Nichols (2008)). They concluded that the *way* determinism is described can influence intuitions about free will and moral responsibility.

Results such as these have led some to think that sometimes people incorrectly interpret that determinism entails "bypassing" conscious agency (Nahmias, Coates, et al., 2006a; Nahmias & Murray, 2010).[4] That is, the processes that result in the action go around one's conscious agency so one's beliefs and desires are not interpreted as playing a role in the production of the action. If conscious agency is bypassed (e.g., one's beliefs, desires, intentions, plans), then in some important respects it is not the person who is acting. If the person is not acting, then the person is not free or morally responsible for those actions. However, if determinism is correctly understood as "going through" conscious agency (i.e., one's mental states are importantly involved in the deterministic causal sequence), then many people have compatibilist intuitions. In these ways, people can sometimes express compatibilist intuitions and sometimes incompatibilist intuitions depending on how they interpret determinism. Abstractly described actions are more likely to encourage an incorrect bypassing understanding of determinism. However, concretely described actions are more likely to discourage a bypassing interpretation by making it more obvious that the person is somehow importantly involved in the production of the action.[5] Hence, bypassing can explain the abstract/concrete difference in free will intuitions.

[4] For an evaluation of the bypassing account, see Rose and Nichols (2013).

[5] There are other explanations of the concrete/abstract difference. For example, the Norm Broken, Agent Responsible account holds that it is clear that often in concrete cases the agent breaks a norm. As such, the person is morally responsible for doing so. In abstract cases, there is no description of a person breaking some norm, so the person is not judged to be morally responsible (Mandelbaum & Ripley, 2012).

Extraversion Predicts Compatibilist Intuitions

These studies contribute importantly to the understanding of folk intuitions about freedom and moral responsibility. But of note, there is a consistent, stable, and substantial dissenting minority in all these studies. For example, A. Feltz, Cokely, and Nadelhoffer (2009) presented participants both of Nichols and Knobe's (2007) High and Low Affect cases. Results revealed remarkable consistency in participants' responses. When asking about whether one can be free and morally responsible, 67% of participants gave incompatibilist friendly responses to both high and low affect conditions whereas only 8% gave a mixed response. Asking about free will revealed largely the same effect. Sixty-two percent of participants gave incompatibilist matched answers and only 9% gave mixed responses. So while Nichols and Knobe's between subjects study suggested that people can be manipulated by affect, Feltz, Cokely, and Nadelhoffer's studies suggested that there is at least some temporal stability in judgments. This finding is consistent with the idea that there may be some relatively stable individual differences in people's judgments about freedom and moral responsibility.

But what could account for the stability in free will judgments? One compelling possibility is that one's personality might be partially responsible for free will and moral responsibility judgments. Many models of personality hold that that personality traits are heritable and are relatively resistant to change over time once one reaches young adulthood (but they can and do change, as well). Moreover, personality traits are associated with differences in how one thinks (e.g., those who are conscientious may take more time deliberating about a difficult problem compared to those who rate themselves lower on conscientiousness) and how one behaves (e.g., those who are high on conscientiousness may have a cleaner room than others) (Funder, 1991, 1995; McCrae & Costa, 1990). Heritable personality traits shape our interests, feelings, and reactions thereby influencing how we interact and relate to the world around us. As such, personality traits may also shape our intuitions about philosophically relevant issues such as freedom and moral responsibility.

Today, results from our research program provide substantial evidence that the global personality trait extraversion predicts how one responds to the combability question. Extraversion is one of the most fundamental dimensions of human personality, and is in some way represented in most major personality models (Lucas, Diener, Grob, Suh, & Shao, 2000). In most of the experimental philosophy research reported here and elsewhere, extraversion is measured via a validated psychological research assessment designed to measure individual differences in the Five Factor model of personality (extraversion, agreeableness, openness to experience, emotional stability, and conscientiousness) (John & Srivastava, 1999). Within the Five Factor model, an extravert is defined as one who is a "communicative, sociable, energetic person who thrives on social contact and who does not regulate tightly his/her emotional reactions" (Akert & Panter, 1988, p. 966). Extraverts enjoy social interaction, find social interaction rewarding, and actively seek out and are motivated to engage in social interactions (Ashton, Lee, & Paunonen, 2002; Lucas et al., 2000).

While there are many features that characterize and help identify what an extravert is, one feature that is particularly relevant for our purposes is the social nature of extraverts. In general, one tendency of extraverts is that they are more sensitive to and understand interpersonal dynamics better than introverts (Akert & Panter, 1988). To illustrate more concretely, extraverts are often better at understanding non-verbal communication compared to introverts. Extraverts also tend to enjoy social interaction, and perhaps partially as a function of that, they have different and socially minded judgment process, interpretations, and memories (Chamorro-Premuzic, Furnham, & Ackerman, 2006; Lucas & Fujita, 2000; Rusting & Larsen, 1997; Zelenski & Larsen, 2002). As we mentioned at the beginning of the chapter, beliefs about freedom and moral responsibility are fundamental elements of how we relate to ourselves and others. In other words, beliefs about freedom and moral responsibility can serve important social functions. So, it stands to reason that extraverts could potentially be differentially influenced by features of cases involving freedom and moral responsibility, and may have systematically different intuitions about those cases compared to introverts.

These general tendencies among extraverts suggest that they may interpret or interact with the materials in the Free Will Scenarios differently from those who are not extraverted. These interpretations or interactions may be especially pronounced when the action that people are asked to make judgments about is something that violates social norms such as killing somebody. In those cases, because of extraverts' judgment tendencies, extraverts may have different reactions to those cases just like they may have different reactions to situations in real life compared to non-extraverts. Just to illustrate, extraverts tend to value having social interactions that are not contentious and are pleasant. If that is right, then when an extravert makes a judgment about somebody who engages a contentious and unpleasant action, extraverts' judgment tendencies may be triggered and override any of the potential excusing conditions for that action (e.g., that the person's action was completely caused by what happened before). Those excusing conditions may have relatively less weight (or go unnoticed) because of the socially aberrant or affective nature of the action they are asked to make judgments about. Rather, extraverts' judgments may be primarily driven by a desire to maintain social balance and harmony. In other words, extraverts may judge a person to be free and morally responsible because holding them to be free and morally responsibly is socially important (for related line of reasoning, see Smilansky (2002)).

Given that there are some good theoretical reasons to suspect that extraversion is related to compatibilist intuitions, it still remains to be seen whether extraversion is in fact related to compatibilist intuitions. Efficiently establishing this relation requires actually conducting some studies to measure the relation between extraversion and compatibilism. Typically, this involves presenting some description of determinism, measuring responses to that description, and then estimating the relation of responses to the global personality trait extraversion. There are a number of ways to describe determinism, a number of different ways to measure responses to those descriptions, and a number of different groups of people whose intuitions could be measured. Having different possible descriptions, response options, and possible samples provide challenges and opportunities. The challenge is that *some* way to describe determinism and measure responses has to be chosen. But there is no widely agreed upon single best way to describe determinism. All ways of describing determinism can be criticized as not capturing, or not communicating, the central aspects of determinism (A. Feltz et al., 2009; J. Turner & Nahmias, 2006). Additionally, one has to choose the prompts and the response options (e.g., yes/no, rating agreement, open-ended responses, etc...), each having their own strengths and weaknesses. Finally, one has to select the relevant group of people. If one wants to know what professional philosophers think about freedom and moral responsibility, it would be a poor strategy to select people who have no training in philosophy. A decision must be made about how best to ensure representative sampling and assessment.

Having multiple descriptions of determinism and ways to respond is also an opportunity. By presenting different groups of people with different descriptions of determinism and offering different methods of responses, we can begin to estimate the extent to which evidence and intuitions converge while mapping out the range and stability of the intuitions that people express about determinism. This is particularly important for predicting intuitions with personality traits because one worry is that a personality trait could be related to just one description of determinism, group of people, or set of response options. If that were the case, then extraversion may not be related to compatibilist intuitions but rather something more specific about one scenario, sample, or handful of response options. If those descriptions and response options converge on the same general pattern across theoretically related assessment instruments and samples, then there is good reason to suspect that people's responses are not only dependent on idiosyncratic features. That pattern of results would converge to provide compelling evidence that some people have deep-seated compatibilist intuitions that are robustly related to their personality.

While theoretical reasons can guide decisions about materials, response options, and target populations to some degree, at one point a choice has to be made. In one of the seminal studies to demonstrate empirically the relation between extraversion and compatibilist intuitions, a psychologically non-reductionistic scenario based on the scenarios used by Nahmias, Coates, et al. (2006a) was chosen to describe determinism (A. Feltz & Cokely, 2009). One reason this scenario was chosen was because it is very similar to the general types of scenarios used to measure compatibilist judgments. In this sense, the scenario was chosen because it was "industry standard"[6] and many if not most experts agreed it represented key features of determinism. The scenario was identical to the *psychologically non-reductionistic* scenario above except that the second paragraph was replaced with the following paragraph:

> *High Affect*
>
> So, once specific earlier events have occurred in a person's life, these events will definitely cause specific later events to occur. For example, one day a person named John decides to kill his wife so that he can marry his lover, and he does it. Once the specific thoughts, desires, and plans occur in John's brain, they will definitely cause his decision to kill his wife.

[6] Just because the scenario was industry standard does not mean it, or the response options associated with it, ought to be industry standard (see Sommers (2010, 2014)). Rather, industry standard used here is just a description of the state of the science at a time.

Fifty-eight participants were asked to rate how much they agreed with the following statements on a 7-point Likert scale (1 = strongly disagree, 4 = neutral, and 7 = strongly agree)[7]:

1. "John's decision to kill his wife was 'up to him.'
2. John decided to kill his wife of his own free will.
3. John is morally responsible for killing his wife."

We will refer to this group of items (1–3) as the *free will questions* since they are some of the standard questions used to assess compatibilist intuitions. As we will see, responses to these three prompts are almost always strongly correlated with one another. Participants were also given the Ten Item Personality Inventory (TIPI) (Gosling et al., 2003), which involves the same set of questions introduced in Chap. 1. We will focus on the two items that make up the extraversion scale in the TIPI. (You can check your scores again from Chap. 1 to see if our results match with your experience with the scenarios.)

Figure 2.1 presents proportions of participants' responses dichotomized according to whether participants agreed (response > 4), were neutral (response = 4), or did not agree (response < 4) to the free will questions:

Replicating Nahmias, Coates, et al. (2006a) study, most people agreed that the person was free and morally responsible. The mean response for

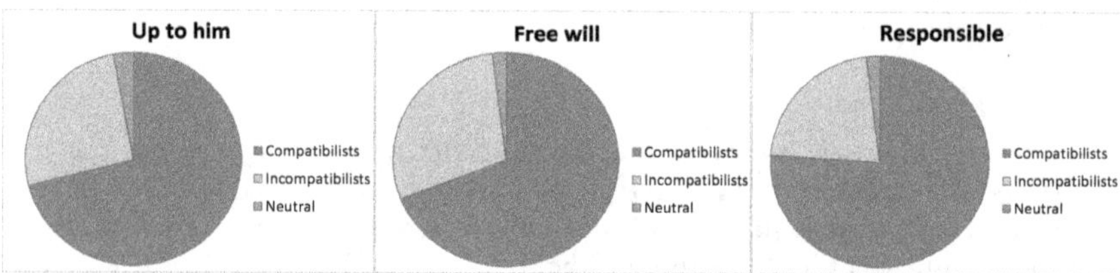

Fig. 2.1 Percent of response to the free will questions

[7] The scale in the original experiment was reversed, where 1 = strongly agree, 4 = neutral, and 7 = strongly disagree. The presentation is changed here to be more consistent with reports from subsequent studies.

all questions was on the agreement side of the scale, suggesting that the mean response was compatibilist friendly.[8]

We estimated the correlations between the three free will questions (questions 1–3 above). These estimates suggested that there were strong relations among the responses to the three questions (*rs* > .7). When items are so strongly correlated with one another it suggests that, to a great extent, they are all likely measuring the same basic thing. So a *composite compatibilism score* was calculated ((answers to the three free will prompts)/3). We then calculated correlations among each of the three free will questions, the composite score, and extraversion. Extraversion was related to each of the 3 free will questions as well as the composite score. In short, as one's extraversion level went up so too did agreement with the free will questions. However, when we calculated the correlations with the other Big Five personality traits, gender, and other general cognitive abilities (e.g., cognitive impulsivity), no other statistically significant correlations were found (see Table 2.1).

To further illustrate these relationships, we divided participants into four groups (i.e., quartiles) depending on how they scored on the extraversion subscale of the TIPI. Then we performed what is sometimes called an "extreme groups analysis" (Cokely, Kelley, & Gilchrist, 2006). That means that we took those who were in the top extraversion quartile and compared those participants to those who were in the bottom extraversion quartile. The reasoning is that in an extreme groups analysis, we should see the differences between those who are highest in the trait of extraversion and those who are the lowest. That is what the extreme groups analysis found (see Fig. 2.2). Those in the bottom extraversion quartile had

Table 2.1 Correlations of participants' responses ($N = 58$)

	Up to him	*Responsible*	*Compatibilism*	*Extraversion*
Free will	.73**	.70**	.90**	.27*
Up to him		.73**	.91**	.38**
Responsible			.90**	.26*
Compatibilism				.34**

* $p < .05$, ** $p < .01$

[8] Up to him: $M = 5.28$, $SD = 2.03$; free will $M = 5.05$, $SD = 2.06$; responsible: $M = 5.65$, $SD = 2.13$.

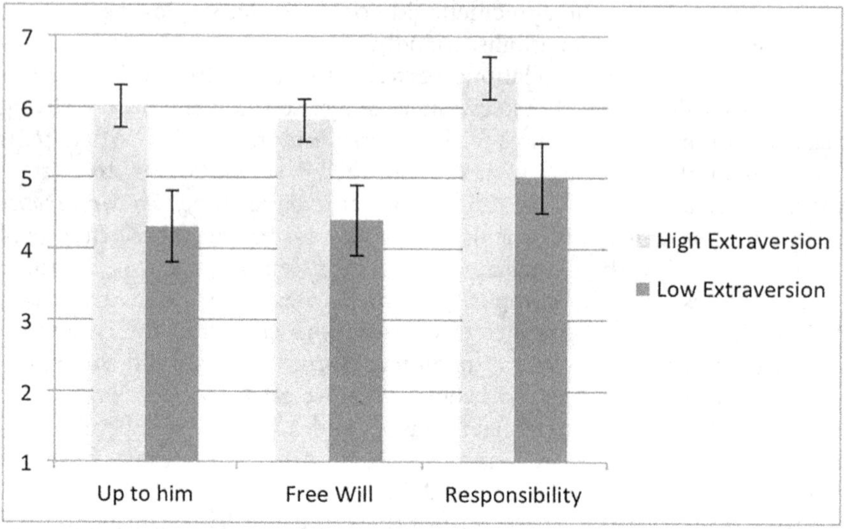

Fig. 2.2 Low (bottom quartile) and high (top quartile) extraversion scores by level of agreement with Up to him, Free will, and Responsibility statements. Error bars represent the standard error of the mean

statistically significant lower, and arguably qualitatively different (i.e., neutral) free will judgments compared to those in the top quartile (*Cohen's d* ranging from .6 to .9).

These results suggest that extraversion was a reliable predictor of some compatibilist intuitions. To summarize, the correlation table presented in Table 2.1 suggested that extraversion was moderately related to free will judgments in the predicted direction. As one is more extraverted, one tends to agree more strongly that the person has free will and is morally responsible. This general pattern was further illustrated by an extreme groups analysis based on extraversion quartiles. And, as we have noted, on at least two of the free will questions, there was arguably a qualitative shift where people who were low in extraversion did not agree that the person was free nor did they agree the action was up to him.

An increasingly important element in psychology, and science more generally, is replication. By some estimates, more than 50% of findings reported in leading journals in medicine, science, genetics, etc., is likely to fail to replicate. If the previous study were the only piece of evidence that extraversion predicts compatibilist intuitions, then we would have limited reason to think that extraversion *generally and robustly* predicts compatibilist intuitions. After all, the results from this one study could have capitalized on chance (Type I error or a statistical false alarm) or may have been related to specific features of the scenarios, participants, or responses options rather than reflecting something deep and enduring about compatibilist intuitions. While the statistical analyses suggest these results were likely to generalize, some caution is merited when interpreting the results of just one study of 50 young adults living in the United States.

Fortunately, the relation between extraversion and compatibilist intuitions has since been replicated many different times in different labs, in different countries, with different and diverse samples, using different descriptions of determinism and response options (Andow & Cova, 2016; E. T. Cokely & Feltz, 2009a; A. Feltz, 2013, 2015a, 2015b; A. Feltz & Cokely, 2008; A. Feltz & Millan, 2015; A. Feltz, Perez, & Harris, 2012b; Nadelhoffer, Kvaran, & Nahmias, 2009; Schulz, Cokely, & Feltz, 2011). There are of course some exceptions and some people who think the exact opposite of what we'd expect, but the strength of the observed relationship as estimated in the seminal study has been generally consistent across studies. Using a technique called meta-analysis, we were also able to rigorously combine the results different, discrete studies. This statistical combination aimed to provide the "best available" estimate of relation between extraversion and compatibilist intuitions. This technique generally helps alleviate many of the worries of findings based on individual studies, such as the risk of capitalizing on chance relations. The best meta-analytic estimate at the time of writing this chapter is that the overall average relation between extraversion and compatibilist intuitions is about .2 (*95% CI* .15–24) (A. Feltz & Cokely, 2019). We have included a figure (sometimes called a Forest Plot) to illustrate the size of the effects for each individual study reviewed (see Fig. 2.3).

The square dot in the middle of the lines for each study represents the sample size of that study—bigger squares mean bigger samples and therefore they are weighted more heavily in the overall estimate of the overall effect size. The lines surrounding the box are confidence intervals. The confidence interval represents an analytic estimate depicting the precision of the estimated interval (e.g., the true size of the effect is very likely to fall somewhere within that range). Theoretically, if all the studies are estimating the same underlying relation (i.e., extraversion's correlation with compatibilist intuitions), then 95% of all studies should have a confidence interval that includes that true value (and 5% should not). This theoretical

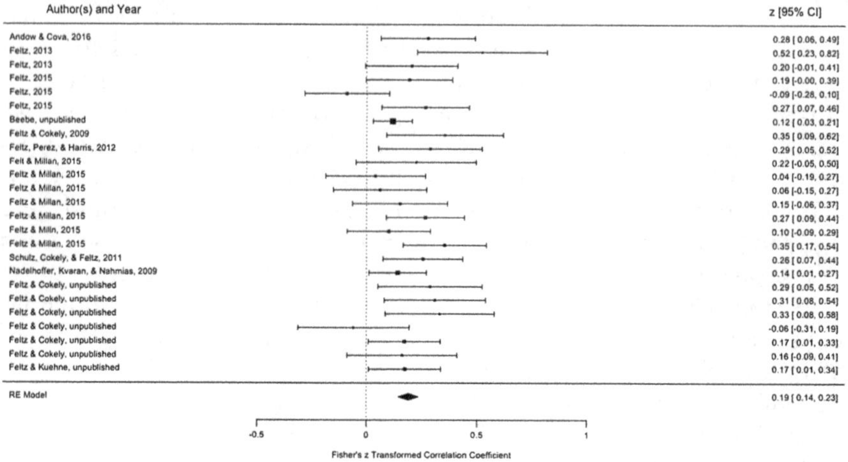

Fig. 2.3 Forest plot of effects in studies about the relation between free will judgments and extraversion

pattern is what we see with the relations between extraversion and compatibilist intuitions. Nearly all the studies have a confidence interval that includes .2, the best estimate of the relation. The diamond at the bottom of the figure represents the overall mean correlation based on the studies analyzed and the edges of the diamond represent the confidence interval for the overall mean correlation.

The result of the meta-analysis suggests that there is a robust, stable relation between extraversion and compatibilist intuitions across labs, testing environments, and sampling techniques. To put the estimated strength of the average relationship between extraversion and compatibilism in perspective, it is about the same as the estimated general strength of the relationship between human weight and sex. Yes, some women weigh more than some men, but most of the time it's a good bet that a random group of men will outweigh a random group of women by a substantial margin, just as a random group of extraverts will on average feel much more strongly about compatibilism than a random group of introverts, regardless of their education, income levels, general cognitive abilities, decision making skills, cultural backgrounds, ages, or gender identities. The effect is strong enough that it implies that most of the time a United States Congress consisting of all extraverts would vote differently than a Congress consisting of all introverts when voting on censures condemning the behaviors of politicians.

Determinism, Fatalism, and Individual Differences

Some theorists have worried that participants do not fully understand the deterministic nature of the scenarios used to measure compatibilist intuitions (A. Feltz et al., 2009; A. Feltz, Tanner, Hoang, Holt, & Muhammad, 2022; J. Turner & Nahmias, 2006). The worry is reasonable because the philosophical sense of determinism is highly technical and nuanced. The notion (i.e., a full definition of the word "determinism") is not likely to be reflected in vernacular and is not likely to be well understood by many non-experts. The difficulty with determinism was partially the impetus to provide people with descriptions of determinism in scenarios rather than just a definition. Given the scenarios approach, it could be relatively easy for some participants to incorrectly interpret or incompletely process the deterministic information presented in the scenarios. For example, J. Turner and Nahmias (2006) and A. Feltz et al. (2009) have commented that the scenarios used by Nichols and Knobe (2007) and Nichols (2004b) may encourage some people to understand the scenarios fatalistically. Fatalism as we will use the term is "the thesis that whatever happens must happen; every event or state of affairs that occurs, must occur, while the nonoccurrence of every event and state of affairs is likewise necessitated" (Bernstein, 2002, p. 65). Nichols and Knobe (2007) and Nichols (2004b) use scenarios that describe an agent as "having to" act a certain way. But actions do not "have to happen" in a deterministic world. If laws of nature or initial conditions had been different, then different actions could have come about. Actions that are fated must happen regardless of the causal nature of the universe, the laws of nature, or any state of the universe (Bernstein, 2002). As such, the language used by Nichols and Knobe may encourage a fatalistic rather than a deterministic understanding of the scenarios. Moreover, many compatibilists believe that fated actions are not done freely. If participants do not appropriately understand the deterministic nature of the scenario, then their responses do little to help illuminate the compatibility question, much less extraversion predicting compatibilist intuitions. Hence, it is difficult but critically important to convey to non-philosophers an accurate notion of determinism that does not encourage an unwanted understanding of the scenarios (e.g., fatalism, indeterminism) if folk intuitions are to help inform answers to the compatibility question (see also Nadelhoffer, Rose, Buckwalter, & Nichols, 2020a).

A. Feltz and Millan (2015) tested whether people can appreciate the difference between fatalism and determinism. In a variety of different scenarios, they found that overall people had different intuitions about some determined v. fated actions.[9] Here's one fatalistic scenario they used:

> *Book:* There is a special book that has all of our decisions and actions truly written in its content. For instance, whenever we are trying to decide what to do, the decision we end up making is completely and truly written in this book. The special book has these events truly written in it lifetimes before the events took place.
>
> So, if the book has an event written in it, the event will definitely occur. For example, one day a person named John decides to kill his wife so that he can marry his lover, and he does it. Once the specific event is truly written in the book, it is impossible for John not to kill his wife.
>
> Assume the book's contents made it impossible for John not to kill his wife. Please rate to what degree you agree with the following statements.

Participants responded on a 7-point scale (1 = strongly disagree, 7 = strongly agree). A separate group of participants also received a low affective version of the fatalism case describing John cheating on his taxes. Participants answered the free will questions along with the following comprehension question "If the universe were re-created with the special book having the same true sentences, John would do the same thing?". Two separate groups of participants received only one of the high and low affect scenarios based on Nahmias, Coates, and Kvaran's determinism scenarios. Overall, participants had different patterns of responses to fated and determined actions (see Table 2.2 for means and standard deviations). Those in the Book scenario had statistically significant, and large, differences in free will judgments compared to those in the Determinism scenario (of note, these studies did not detect a reliable difference of affect and affect did not interact with conditions). In these studies, extraversion was related to responses to the determinism scenario but not to the fatalism scenario. To provide evidence for that relation, they conducted a hierarchical linear regression that first entered (1) high/low affect and (2) Sex

[9] Feltz and Millan used other fatalistic scenarios including God's foreknowledge and a crystal ball that infallibly sees the future. All scenarios revealed the same difference with a typical determinism scenario.

Table 2.2 Means and standard deviations for Fatalism and Determinism scenarios

	Fatalism (N = 76)	*Determinism (N = 55)*
Up to	$M = 2.8$, $SD = 2.05$	$M = 5.38$, $SD = 1.85$
Free will	$M = 2.88$, $SD = 2.12$	$M = 5.57$, $SD = 1.54$
Responsibility	$M = 4.0$, $SD = 2.15$	$M = 5.86$, $SD = 1.67$

to control for variance that could be associated with those variables. Then they entered the extraversion score into the model. The overall model significantly predicted compatibilist judgments $F(3, 51) = 8.64$, $p < .001$, $R^2 = .34$. Even after controlling for high/low affect and sex, extraversion continued to predict unique variance, $\beta = .35$, $t = 1.97$, $p = .05$, $R^2_{change} = .05$. However, when they performed the same analyses using responses to the Fatalism scenario, they did not find a reliable relation between extraversion and responses to the Fatalism scenario ($t < 1$).

These results were replicated in a subsequent study with the same materials using a within-subject design (i.e., the same people made judgments about both vignettes but at different times). Participants were given one pair of either fatalism and determinism high effect or fatalism and determinism low effect, counterbalanced for order. Participants answered the same free will questions used in previous studies. A mixed-model ANOVA with the two composites scores as within-subjects factors and order of presentation and affect as between subject factors revealed the main effect of Fatalism (M = 3.85, $SD = 2.24$) and Determinism ($M = 5.58$, $SD = 1.55$): $F(1, 63) = 41.03$, $p < .001$, $\eta_p^2 = .39$. Neither order of presentation ($F(1, 63) = 1.89$, $p = .28$, $\eta_p^2 = .02$) nor affect ($F(1, 63) = 1.16$, $p = .29$, $\eta_p^2 = .02$) interacted with judgments. A hierarchical linear regression was constructed to test extraversion's relation to the compatibilist composite score after controlling for affect, order, and sex. The full model was a near significant predictor of compatibilist judgments $F(3, 67) = 2.23$, $p = .065$, $R^2 = .11$. After controlling for affect, order, and sex, extraversion predicted unique variance in the compatibilist composite score, $\beta = .12$, $t = 2.0$, $p = .05$, $R^2_{change} = .06$. Once again, extraversion did not predict judgments in the fatalism scenario ($t < 1$). These results have also been replicated by a different research group (Andow & Cova, 2016).

Feltz and Millan also reported evidence from a third experiment including the scenarios from Nichols and Knobe (2007). Since there was no reliable effect of affect, the third experiment used only high affect cases. Participants received Book along with Nichols and Knobe's High Affect scenario. Participants responded to the free will questions. A mixed-model ANOVA with the two composites scores as within-subjects factors and order of presentation as between subject factors revealed an overall main effect between Book ($M = 3.75$, $SD = 2.21$) and the determinism scenario ($M = 5.21$, $SD = 1.92$) $F(1, 92) = 40.37$, $p < .001$, $\eta_p^2 = .31$. A hierarchical linear regression with (1) order, and (2) sex and (3) extraversion was a significant predictor of compatibilist judgments $F(3, 89) = 2.81$, $p = .03$, $R^2 = .10$. After controlling for order of presentation and sex, extraversion continued to predict unique variance in the compatibilist composite score, $\beta = .16$, $t = 2.14$, $p = .04$, $R^2_{change} = .05$. Extraversion did not reliably predict unique variance for the fatalism composite score ($t = 1.03$, $p = .31$).

This series of studies suggests everyday intuitions about fated and determined actions tend to be different. Overall, people are more likely to think that one is freer in a determined scenario than in a fated scenario. These studies also suggest that there is something unique about determined actions that is responsible for the relation to extraversion since the relation between extraversion and fated actions was not found. Therefore, we have some evidence that extraversion is specifically related to compatibilist intuitions and not to similar but conceptually distinct threats to freedom and moral responsibility such as fatalism. In the next section, we will see that this pattern of unique predictive ability of extraversion persists for another threat to freedom and moral responsibility—i.e., manipulation.

Free Will and Manipulation

The experimental exploration of free will has largely centered on directly assessing the compatibility question. Direct assessments of the compatibility question probe specific intuitions about the relations of determinism, freedom, and moral responsibility. Tamler Sommers (2010, 2014) argues that this general approach is a mistake—or is at least incomplete. While assessing and documenting intuitions about the compatibility question can be interesting, that's not what philosophers do. Rather, in typical philosophical practice, philosophers give reasons for thinking that some claims are true. These reasons typically are incorporated into an argument for some conclusion. Typically, experimental philosophy does not assess the arguments or reasons that philosophers give for thinking some philosophical claims are true. Most experimental philosophy instead assesses intuitions about the conclusions of those arguments.

To illustrate, one way that experimental philosophers have taken the results of studies to be important is that they help situate argumentative burdens. That is, those results can indicate which views are counter-intuitive. Give the counter-intuitiveness of the positions, those philosophers have to offer additional reasons why their views are correct (or why those with views consistent with the dominant view are wrong) while those with positions supported by folk intuitions do not. In the free will debate, that might suggest that those with incompatibilist intuitions have some additional argumentative burden that compatibilists do not have because compatibilism seems supported by folk intuitions. However, Sommers argues:

> [I]ncompatibilists can accept that they have this 'argumentative burden' but claim that they have discharged it with, well, arguments. After all, van Inwagen's 'consequence argument', Strawson's 'basic argument', and Pereboom's 'four case argument', to name just a few, are designed to precisely lead the reader to the conclusion that determinism precludes free will and moral responsibility... It would seem that in order to truly test the plausibility of the incompatibilist position, we need to examine the intuitions supporting the premises and principles of their argument... (2010, pp. 205–206)

Sommers goes on to argue that starting with a conclusion being correct just begs the question—something that philosophers seldom do and very few think this is a good argument form. Rather, philosophers may have some view and then they offer arguments for those views. Those arguments don't beg the question, so claiming that some view is counter-intuitive simply does not appreciate something important—namely, the reasons given for that incompatibilist conclusion.

Rather than directly assessing the intuitiveness of conclusions of arguments, Sommers suggests testing the reasons that philosophers give for thinking those conclusions are true. For example, Sommers suggests testing the intuitiveness of the four case argument for incompatibilism (Pereboom, 2001). The four case argument is at least in part designed to call into question a key claim made by many compatibilists. Compatibilists often claim that if one's psychological states are related in the right way with the production of an action, then in those conditions one can be free and morally responsible in deterministic environments. Some of these psychological states involve not being constrained to act (Hume, Selby-Bigge, & Nidditch, 1978), having desires play the right causal role in the production of an action (Ayer, 1952), having the action issue from the character of the individual (Hume et al., 1978), having first-order desires matching with second-order desires (Frankfurt, 1971), being moderately reasons responsive (Fischer & Ravizza, 1998), and being able to appreciate and act from moral reasons (Wallace, 1994). Following Pereboom, we call these kinds of psychological states "causal integrationist" conditions.[10]

As will become a common theme in this book about philosophical concepts in general, there is lots of philosophical debate about the causal integrationist conditions (e.g., which conditions are the right ones, are they really sufficient, etc…). One argument that has been offered to suggest that the causal integrationist conditions are not sufficient is Pereboom's "four case argument." The general goal of the four case argument is to start with a scenario where one is obviously not free and morally responsible. Then, through modest changes to the scenario that do not change the judgment about the protagonist's freedom and moral responsibility, end up with a case that describes determinism. The actual sequence of scenarios involves many of the prominent causal integrationist conditions and the following: (1) A completely manipulated person, (2) A programmed person, (3) An indoctrinated person, and then finally (4) A determined person. If Pereboom is right, then most people would judge the person in 1 as not free or morally responsible. Pereboom then claims that "an agent's non-responsibility under covert manipulation generalizes to the ordinary [deterministic] situation" (2001, p. 112). Pereboom predicts: "If I am right, it will turn out that no relevant difference can be found among these cases that would justify denying responsibility under covert manipulation while affirming it in ordinary deterministic circumstances, and that this would force an incompatibilist conclusion." Sripada (2012) has formalized the argument as the following:

1. A manipulated person is not free.
2. There is no relevant difference between a manipulated person and a person in a deterministic world.

[10] Pereboom realizes that most compatibilists do not think that these conditions are sufficient.

3. If there is no difference between a manipulated person and an agent in a deterministic world, then a person in a deterministic world is not free.
4. Therefore, a person in a deterministic world is not free. (C. S. Sripada, 2012)

As Sripada's formalization highlights, the intuitions about the four case argument are support for *premises* in an argument. In particular, the four case argument is supposed to support premise 1 and 2. After that, and along with some basic logic, we are supposed to support the incompatibilist conclusion in 4. Of course, it is an open question about how widely shared Pereboom's intuitions about the four case argument are. But perhaps more importantly, are these intuitions systematically related to personality?

To test the extent to which people shared intuitions consistent with Pereboom's prediction about the four case argument and whether personality predicts those intuitions, we modified cases that were used in the four case argument. Special attention was paid to making the cases accessible and understandable to most people (fancy jargon, like that expressed in the causal integrationist conditions above, was not used), and the same general approach Sripada (2012) used was adopted. Following the stipulations of Pereboom's four case Argument, each case involves a person who arguably meets at least some of the causal integrationist conditions. Participants were assigned to one of two conditions. The first condition, people received all four cases in order. In the other condition, participants only received the final case that is meant to approximate a scenario where determinism is true. Everyone read the following paragraph (A. Feltz, 2013):

> One day Bill sees a woman named Mrs. White as she is jogging in the park. Bill hates this woman, and deliberates about what to do. After weighing his options, Bill decides he should kill her. Bill's mind is not clouded by rage or other extreme emotions. Rather, Bill thinks clearly and carefully about his own desires and values, and only then makes a decision. After he kills Mrs. White, Bill reflects on his action. He wholeheartedly endorses what he has done. BUT, there is more you need to know about Bill, and how he came to be the person he is now...

Then, participants read the following depending on which condition they were assigned (all four in one condition versus only the last paragraph in the other condition):

Intentional Direct Manipulation

Bill is essentially a normal man, but he was created by neuroscientists who directly manipulate all of his decisions. The neuroscientists manipulate Bill to make decisions that almost always benefit him. The neuroscientists implant in Bill a desire to kill Mrs. White. He is able to regulate his behavior by moral reasoning and act differently in different situations with different reasons, but in the present circumstances, the desire to kill Mrs. White is stronger than any competing desire. As a result of the neuroscientists' implanting in Bill the desire to kill Mrs. White, Bill decides to kill Mrs. White and does it. Reflecting on the action afterward, Bill identifies with the desire to kill Mrs. White and the resulting action.

Intentional Indirect Manipulation

Bill is essentially a normal man, but he was created by neuroscientists who do not control him directly, but have programmed his genes so that he makes decisions that almost always benefit him. The neuroscientists program Bill to have a desire to kill Mrs. White. He is able to regulate his behavior by moral reasoning and act differently in different situations with different reasons, but in the present circumstances, the desire to kill Mrs. White is stronger than any competing desire. As a result of the neuroscientists' programming Bill to have the desire to kill Mrs. White, Bill decides to kill Mrs. White and does it. Reflecting on the action afterward, Bill identifies with the desire to kill Mrs. White and the resulting action.

Culture

Bill is essentially a normal man, but he was extensively trained by his community to make decisions that almost always benefit him. He could not have prevented this extensive training, and it is ingrained in him. The extensive training generates in Bill a desire to kill Mrs. White. He is able to regulate his behavior by moral reasoning and act differently in different situations with different reasons, but in the present circumstances, the desire to kill Mrs. White is stronger than any competing desire. As a result of the extensive training generating in Bill the desire to kill Mrs. White, Bill decides to kill Mrs. White and does it. Reflecting on the action afterward, Bill identifies with the desire to kill Mrs. White and the resulting action.

Determinism

Bill is a normal man raised under normal circumstances. Every decision that Bill makes is completely caused by his genes and his cultural environment. He is able to regulate his behavior by moral reasoning and act differently in different situations with different reasons, but in the present circumstances, the desire to kill Mrs. White is stronger than any competing desire. As a result of his genes and his cultural environment generating in Bill the desire to kill Mrs. White, Bill decides to kill Mrs. White and does it. Reflecting on the action afterward, Bill identifies with the desire to kill Mrs. White and the resulting action.

Then participants responded to each of the following prompts (1 = strongly disagree, 7 = strongly agree).

1. "Bill kills Mrs. White of his own free will.
2. Bill's killing of Mrs. White is 'up to him.'
3. Bill is morally responsible for killing Mrs. White.
4. Bill is blameworthy for killing Mrs. White.
5. Bill deserves to be punished for killing Mrs. White.
6. Bill should be prevented from killing Mrs. White."

Consistent with the inter-relations of free will questions noted above, the responses to 1–6 were related to one another. So, for ease of analysis, a composite score that was the mean of 1–6 was calculated. The responses in the four case condition were inconsistent with Pereboom's prediction. There was a statistically significant difference among the conditions where people judged the person to be not free or morally responsible in the Intentional Direct Manipulation case ($M = 3.88$, $SD = 1.6$). But participants were in increasing agreement that the person in the subsequent scenarios was free and morally responsible: Intentional Indirect Manipulation ($M = 4.28$, $SD = 1.54$), Culture ($M = 4.91$, $SD = 1.61$), and Determinism ($M = 5.71$, $SD = 1.61$).

But what about personality's relation to the judgments in the four case condition and the single case condition? As in the studies reviewed above, participants completed the TIPI (Gosling et al., 2003). The relation of personality with the mean of judgments to prompts 1–6 was assessed. In the single Case condition, once again the relation of extraversion with compatibilist judgments was replicated $r\,(46) = .48$, $p = .001$. However, extraversion was not related to compatibilist judgments to Determinism in the four case condition: $r\,(40) = -.17$, $p = .31$. Rather, in the four case condition a different personality trait was related to judgments. Emotional stability was positively related to judgments: Intentional Direct Manipulation $r\,(40) = .27$, $p = .09$, Intentional Indirect Manipulation $r\,(40) = .31$, $p = .05$, Culture $r\,(40) = .32$, $p = .04$, Determinism $r\,(40) = .34$,

$p = .03$. No other personality trait was related to judgments in the four case condition (all other p's > .13). This study thereby replicated the effect of extraversion for compatibilist intuitions (i.e., in the single Case condition) yet found another relation of a global personality trait with a series of studies involving manipulation. It appeared that those who were emotionally less stable were more likely to be influenced by the manipulation scenarios that occurred earlier in the series, which reduced their strength of agreement with a person's freedom and moral responsibility in later cases.

The effect of emotional stability was surprising and unanticipated. Sound and responsible scientific practice necessitates a replication in cases like these. A common strategy in such cases is to develop a study to "replicate and extend" (e.g., doing the important work of confirming the unexpected finding but doing so in the context of what is generally more prized by scientists—advancing a frontier). So that's what was done with one important modification to the four cases. Mele (2006) speculates that people may be responsive to the nature of the manipulator in the four cases presented above. In particular, in each of the two manipulator scenarios, there was intentionality behind the actions of the protagonist (e.g., in the Direct Manipulation scenario, the neuroscientists intentionally manipulate all of Bill's actions). In the other scenarios, there was no intentionality to the influence on Bill's actions.

To help address this issue in the replication study, the manipulation in the first two cases was changed to be the result of a brain tumor that either completely manipulates all Bill's actions or programs him to behave as he does (e.g., "Bill is essentially a normal man, but he has a *brain tumor* that directly manipulates..."). It is plausible to think that the brain tumor does not manipulate Bill intentionally, so this change should satisfy Mele's concern about the original four cases. Again, participants in this study were either given all four cases or were in the single case condition.

Results of the new replication and extension study were again contrary to Pereboom's prediction and the means for each case was roughly similar to those observed in the first four case study. The same correlations of personality with free will judgments was also observed: extraversion was positively related to the single case Determinism composite score, $r(90) = .20$, $p = .056$; however, this relation was again not found in the four case scenario: $r(89) = .03$, $p = .80$. Nevertheless, emotional stability was again related to intuitions in three of the four cases in the revised four case scenario: Non-Intentional Direct Manipulation $r(89) = .33$, $p = .001$,

Non-Intentional Indirect Manipulation r (89) = .32, p = .003, Culture r (89) = .08, p = .48, Determinism r (89) = .22, p = .04.

The series of manipulation studies provided additional converging evidence for the relation of extraversion and compatibilist intuitions. In both of the single case Determinism scenarios, extraversion predicted compatibilist responses. This relation was eliminated in the four case condition, again indicating that extraversion predicts specific kinds of free will and moral responsibility judgments. Indeed, in the four case scenarios, extraversion did not predict compatibilist intuitions in the determinism scenario. Rather, a different general personality trait, emotional stability, predicted intuitions in the four case scenarios.

The differential predictive ability of different personality traits should be expected since it simply is not the case that any single personality trait (e.g., extraversion) should predict *all* attitudes about freedom and moral responsibility. That would be similar to saying that a single personality trait should predict all attitudes about other broad domains (e.g., friendship, justice). Judgments about freedom and moral responsibility are diverse and complicated, and therefore are likely to be most efficiently predicted with a variety of personality traits. However, for our purposes, what is important is to identify the extent to which any one personality trait can reliably and robustly predict judgments about freedom and moral responsibility.

Free Will, Individual Differences, and Language

All the studies reviewed thus far have been conducted in English with samples drawn from the United States. This leaves open the following question: Does personality predict intuitions in different languages and cross-culturally? There are some reasons to think that personality would predict across cultures and demographics, as well as some reasons to think that personality would not. One reason personality may not predict intuitions about free will and moral responsibility is that there are some cross-cultural differences in cognition. To take one example, East Asian people tend to be more holistic than Westerners (Nisbett, Peng, Choi, & Norenzayan, 2011). Westerners have a greater tendency to attribute the causes of actions to internal mental states of the person whereas Easterners tend to focus more on the context in which the actions originated (Morris, Nisbett, & Pent, 1995). In this sense, East Asians may take the deterministic (or other contextual) factors of a person's action more seriously compared to Westerners who tend to focus on internal mental states (e.g., thoughts, desires, and beliefs). So, the different focus may eliminate the relation of personality to freedom and moral responsibility intuitions. A second reason is that there are some instances of cultural variability in philosophically relevant intuitions. Some semantic intuitions (Machery, Mallon, Nichols, & Stich, 2004) and some epistemic intuitions (J. Weinberg, Nichols, & Stich, 2001) vary as a function of culture, although some of these effects have been somewhat difficult to replicate (Lam, 2010; Seyedsayamodst, 2015; Ziółkowsk, Wiegmann, & Horvath, 2023), while others have successfully replicated (e.g., Gödel cases) (Dongen, Colombo, Romero, & Sprenger, 2020), see Cova et al. (2021)).[11] So, given the variation, there is some reason to think that the relation to personality might not be reliably present in other cultures.

However, there are some good reasons to think that personality would predict free will and moral responsibility intuitions cross-culturally. Sarkissian et al. (2010) examined possible cross-cultural differences in intuitions about free will and moral responsibility in India, Hong Kong, Columbia, and the United States. They gave participants a typical scenario describing determinism like those presented above. They then asked the abstractly framed question "is it possible for a person to be fully morally responsible for their actions?". There were no significant cross-cultural differences in responses. Consistent with the concrete/abstract different in free will intuitions, the majority of participants across the different regions of the world thought that moral responsibility was not compatible with determinism in this abstract frame (63–75%).

If Sarkissian and colleagues are correct that some intuitions about free will are a cultural universal, then judgments about concretely described individuals would likely be compatibilist friendly cross-culturally (reflecting the abstract/concrete difference).[12] Moreover, the relation of extraversion to judgments of freedom and moral responsibility is likely to persist across cultures because there is little difference in personality between people in the United States and South America, Western Europe, and Southern Europe (Schmitt et al., 2007).

[11] For more on cross-cultural studies in experimental philosophy, see the Geography of Philosophy project here https://www.geographyofphilosophy.com/.

[12] Sarkissian et al. (2010) translated their materials that employed abstractly framed questions into Spanish but did not report those translations.

We will review three studies suggesting that extraversion is likely to predict compatibilist judgments cross-culturally. In one of our studies (Schulz et al., 2011), we found that extraversion predicts compatibilist intuitions in a sample of adults of various ages and education residing in Germany. Participants were given the following determinism scenario (in German):

> Most respected neuroscientists are convinced that eventually we will figure out exactly how all of our decisions and actions are entirely caused. For instance, they think that whenever we are trying to decide what to do, the decision we end up making is completely caused by the specific chemical reactions and neural processes occurring in our brains. The neuroscientists are also convinced that these chemical reactions and neural processes are completely caused by our current situation and the earlier events in our lives, and that these earlier events were also completely caused by even earlier events, eventually going all the way back to events that occurred before we were born.
>
> So, if these neuroscientists are right, then once specific earlier events have occurred in a person's life, these events will definitely cause specific later events to occur. For instance, once specific chemical reactions and neural processes occur in the person's brain, they will definitely cause the person to make the specific decision he or she makes. So, once specific earlier events have occurred in a person's life, these events will definitely cause specific later events to occur.
>
> For example, one day a person named John decides to kill a shop owner, because he needs money and does it. Once the specific thoughts, desires, and plans occur in John's mind, they will definitely cause his decision to kill a shop owner.

After reading this scenario, participants were given the free will questions.

Responses to the free will questions were highly correlated (.48, .52, .67; $p < 0.05$ for all) so a composite free will score was calculated for the free will questions. In this case, participants responded to a more detailed measure of extraversion from the NEO-PI-R (Costa & McCrae, 1992, German version: John et al., 2004). On this measure, as is consistent with personality theory in general, the global trait extraversion is constituted by smaller factors called facets. One of these facets of extraversion is warmth.

In this study, only warmth was related to compatibilist judgments explaining about 6% ($p = 0.005$) of the variance in compatibilist judgments.

Two more recent studies we conducted also suggest that the relation of extraversion to compatibilist judgments persists in Spanish. In the first recent study, 129 participants were recruited from Amazon's Mechanical Turk. All materials used to recruit participants were in Spanish. We included some data quality control measures. Nine participants were excluded for not completing the survey and four were excluded for requesting that their answers not be used. Since the materials were in Spanish, we wanted to have some indication that the participants spoke Spanish, so we included a comprehension test in Spanish (see Appendix to this chapter). Forty-five were excluded for incorrectly answering at least one of the Spanish comprehension questions. Seventy-one participants remained for analyses. Fifty (70%) identified as male.

Participants were randomly assigned to receive the Spanish version of only one of either the Determinism High Affect or Determinism Low Affect scenarios (see the Appendix for the Spanish version). One translator fluent in Spanish and English translated the English version of the scenarios into Spanish. A separate translator fluent in Spanish and English translated the scenarios back from Spanish into English.[13] There were no substantial disagreements between translators and there were few discrepancies between translations. All discrepancies were negotiated until both translators agreed. Participants also received a Spanish version of the free will questions. Participants could rate their level of agreement from 1 (strongly disagree) to 7 (strongly agree) to each of the three prompts. Participants also received the Spanish version of the Ten Item Personality Inventory (Gosling et al., 2003).

Responses to free will questions had strong internal consistency (*Cronbach's Alpha* = .89). To simplify analyses, a compatibilist composite score (mean of response to 1–3) was calculated. To test for differences between Determinism High Affect and Determinism Low Affect, an analysis of variance (ANOVA) was conducted with the compatibilist composite score as the dependent variable and condition as the independent variable. Consistent with the meta-analysis discussed above (A. Feltz & Cova, 2014), there was not a significant difference between Determinism High Affect ($M = 5.6$, $SD = 1.45$) and Determinism Low Affect ($M =$

[13] We would like to thank Melissa Millan and Stephanie Samayoa for help in translating and refining these materials.

5.75, $SD = 1.79$) ($F < 1$, $\eta_p^2 = .01$). To increase statistical power, Determinism High Affect and Determinism Low Affect were analyzed together to estimate the relation of compatibilist intuitions with the global personality trait extraversion. As predicted, there was a significant, positive correlation between the compatibilist composite score and extraversion, $r(71) = .28$, $p = .02$. Extraversion was the only personality trait significantly correlated with compatibilist judgments (*rs* between -.12 and .18, $ps > .14$).

Results from the first Spanish language experiment were consistent with results from previous studies using English speakers and provided evidence that the new materials were reliable Spanish language instruments. First, people had overall compatibilist friendly judgments. Second, there was no reliable difference between the high and low affect versions. Third, and importantly, the global, heritable personality trait extraversion predicted compatibilist judgments.

While the consistency of the Spanish language results was predicted based on past theory and evidence, it is still desirable to replicate and extend these findings. The second experiment also assessed the extent to which Spanish speakers could discriminate between determined and fated actions. Specifically, in the second new experiment, 141 participants were recruited from Amazon's Mechanical Turk using the same Spanish language recruitment materials. Participants were excluded because they requested that their answers not be used ($N = 6$) or did not complete the survey ($N = 13$). Forty-seven participants were excluded for failing at least one of the Spanish comprehension questions. Forty-two (56%) identified as male.

The same Spanish language determinism scenarios were used in a new study. In addition, participants received a Spanish version of a scenario that described fatalism (the Book scenario from earlier in this chapter). As in the first experiment, one coder fluent in Spanish and English translated the English version of Book into Spanish. A separate coder fluent in Spanish and English translated the Spanish version back into English (see the Appendix for Spanish language versions). There were only minor disagreements about translations that were resolved with discussion. Participants also received the Spanish language version of the free will questions. Two groups of participants were randomly assigned to either the high affect or low affect conditions. Those in the high affect condition received the Determinism High Affect and Book High Affect ($N = 38$), counterbalanced for order. A separate group received the Determinism Low Affect and Book Low Affect ($N = 37$), counterbalanced for order.

Participants received the appropriate versions of the free will questions. Participants also responded to the Spanish version of the Ten Item Personality Inventory.

Analyses of the second Spanish language study revealed strong internal consistency of judgments about the determinism (*Cronbach's alpha* = .90) and the fatalism scenarios (*Cronbach's alpha* = .91). Compatibilism and Fatalism composite scores were calculated (mean of responses to 1–3). Replicating previous research, a mixed-model repeated measures ANOVA with order and affect as between participant factors and Compatibilism and Fatalism composites scores as within participant factors revealed a large, significant difference between Fatalism (M = 4.1, SD = 2.21) and Compatibilism composite scores (M = 5.59, SD = 1.72), $F(1, 71) = 32.22$, $p < .001$, $\eta_p^2 = 32$. Neither affect ($F < 1$) nor order ($F < 1$) interacted with judgments. To increase statistical power, responses to high and low affect scenarios were combined for subsequent analyses. Extraversion was again related to compatibilist judgments, r (75) = .30, p = .008. Extraversion was the only member of the Big Five reliably related to compatibilist judgments (*rs* < .18, *ps* > .13). Somewhat surprisingly, extraversion was also related to the Fatalism composite score r (75) = .37, p = .001. Openness to experiences was also related to the Fatalism composite score r (75) = .31, p = .007. None of the other three Big Five personality traits were related to the Fatalism composite score (*rs* < .06, *ps* > .6). To test whether extraversion and openness to experience were independent predictors of agreement with the Fatalism composite score, a multiple linear regression with extraversion and openness to experience as predictor variables of the Fatalism composite score was constructed. The full model was a significant predictor of the Fatalism composite score: $F(2, 72) = 8.07$, $p = .001$, $R^2 = .18$. Both extraversion ($\beta = .32$, $t(72) = 2.78$, $p = .007$) and openness to experience ($\beta = .22$, $t(72) = 2.02$, $p = .05$) were independent predictors. Since the relations of personality with the Fatalism composite score were not predicted, they should be interpreted with caution even though the correlations and regression model for extraversion remained significant after a conservative Bonferroni correction ($p = .01$).

Results from these two new Spanish language experiments suggested that the Spanish versions of Determinism and Book assessments accorded with studies using English instruments. Once again, most people had compatibilist friendly intuitions and extraversion predicted compatibilist judgments. Overall, these three studies indicate that extraversion is likely to be a robust predictor of compatibilist judgments across many different languages and cultures.

The Importance of the Individual

So far, we have reviewed a large array of studies suggesting that personality predicts some intuitions about free will and moral responsibility. The evidence we have presented cautions against any sweeping explanation of intuitions about freedom and moral responsibility that does not take into account individual differences. These studies serve as a fine contemporary illustration of why one should not only use overall means to infer actual cognitive processes (Estes, 1956; Molenaar & Campbell, 2009). To illustrate, we will use the affective performance error model to demonstrate the problems associated with not taking into account individual differences when generating models or accounts of the proximal cognitive process involved in philosophically relevant (or other) intuitions. The affective performance error model holds that affective responses generate more compatibilist intuitions. That is, on one natural reading of the affective performance error model, the *same* individual can on one occasion have compatibilist intuitions, yet on another occasion can have incompatibilist intuitions as a function of the affect.[14] The existence of individual differences suggests that folk intuitions about freedom and moral responsibility are likely to be much more stable, yet varied, than this model suggests. To hold that "the folk," as a monolithic entity, engage in an affective performance error would thereby not be accurate. Introverts (or some other identifiable subgroup) may be altogether unaffected by the affective content of scenarios, whereas extraverts (or some other identifiable subgroup) may be influenced by the affective content making their already compatibilist intuitions even stronger. Taking the overall mean response of extraverts and introverts could give the impression that affect influences people overall—or that affect doesn't influence people very much. But that would be inaccurate or incomplete because only one group is reliably influenced by affect. More nuanced accounts are therefore required to capture the variability or boundaries in the processes that generate these intuitions, which are required to have a descriptively accurate understanding of folk intuitions about free will and moral responsibility (or other philosophically relevant intuitions).

[14] Knobe has recently recanted his claim that affect influences people to have more compatibilist intuitions (Knobe, 2014).

To reiterate, one reason why extraversion might be related to compatibilist judgments is that extraverts, compared to introverts, interpret scenarios differently, enjoy social interactions, are more socially motivated, and less tightly regulate their emotional reactions. For these reasons, extraverts may be more likely to hold a person morally responsible especially for bad actions with high affective content in deterministic scenarios. If extraverts have these tendencies, then there is a straightforward prediction about affect: to the extent that the affect is increased, extraverts should be disproportionately influenced by affect compared to introverts. Introverts' judgments should remain relatively stable across scenarios that vary affect whereas extraverts' compatibilist intuitions should be stronger. In other words, extraversion should *moderate* the relation of affect to free will and moral responsibility judgments.

To test this hypothesis, we conducted a preliminary experiment. One hundred and forty-five participants were recruited from Amazon's Mechanical Turk. Four participants were excluded from analyses for not completing the survey or for requesting that their answers not be used. Fourteen participants were excluded from analyses for failing a comprehension question. Sixty-five (51%) participants identified as male. Ages ranged from 18–65, $M = 31.92$, $SD = 12.01$.

We assigned participants to only one of the Low or High Affect Determinism cases. We then gave participants the free will questions along with a standard comprehension question. Participants then completed the Ten Item Personality Inventory (Gosling et al., 2003). Basic demographic information was collected after completing the Ten Item Personality Inventory.

Responses to the free will questions showed strong internal consistency (*Cronbach's* $\alpha = .93$), so a composite free will score was calculated to simplify analyses (mean of answers to 1–3). Replicating Nichols and Knobe's results, an analysis of variance showed that people given the high affect scenario had stronger compatibilist judgments ($M = 4.53$, $SD = 2.11$) than those given the low affect scenario ($M = 3.7$, $SD = 2.36$) $F(1, 125) = 4.29$, $p = .04$, $\eta_p^2 = .03$. However, consistent with the previous meta-analytic results, the effect size was small (about 1% of the variance) (A. Feltz & Cova, 2014). Additionally, replicating previous studies, extraversion was associated with compatibilists responses in the high affect case ($N = 65$), r

$= .32$, $p = .009$; however, extraversion had no reliable relation to judgments in the low affect case ($N = 62$), $r = -.06$, $p = .65$.

Our main concern was whether extraversion moderated or interacted with the relation of affect and compatibilist judgments. An interaction in the statistical sense implies non-linear relations like 1 + 1 = 3 (e.g., dieting may help you lose 5lbs, exercise may help you lose 5lbs but if you diet and exercise you may lose 7lbs or 15lbs). In short, moderation and statistical interaction modeling is used when estimating the extent to which the whole is more (or less) than the sum of its parts. Because the correlations in this study suggested moderation (i.e., correlation in one but not the other case), a hierarchical linear regression was conducted to formally estimate moderation. Extraversion scores were centered and an interaction term was calculated (extraversion score * high/low affect). Extraversion, high/low affect, and the interaction term were entered in that order in different steps of a linear regression. Extraversion ($b = .93$, $SE_b = .41$, $\beta = 1.39$, $t = 2.27$, $p = .03$) and high/low affect (($b = .81$, $SE_b = .39$, $\beta = .18$, $t = 2.07$, $p = .04$) were both significantly associated with compatibilist judgments in the regression. Crucially for our purposes, the interaction was also statistically significant ($b = .24$, $SE_b = .12$, $\beta = 1.28$, $t = 2.1$, $p = .04$; $R^2_{change} = .03$, $F(1, 123) = 4.27$, $p = .04$). Simple slopes were tested, and low extraversion (-1 standard deviation) was not associated with compatibilist judgments ($b = .01$, $SE_b = .55$, $t = .02$, $p = .98$, 95% CI = -1.08 – 1.11). Moderate (mean) ($b = -.81$, $SE_b = .39$, $t = 2.07$, $p = .04$, 95% CI -1.58 – -.03) and high (+1 standard deviation) ($b = -1.63$, $SE_b = .55$, $t = 2.95$, $p = .003$, 95% CI = -2.73 – -0.54) extraversion were related to compatibilist judgments. Indeed, those who were the most strongly extraverted had the strongest compatibilist intuitions in high affect and were the most influenced by the change in affective content.

To illustrate, Fig. 2.4 visually represents the data. For simplicity, the graphs invites you to imagine that there are three groups of people. There is the person who is of "average" extraversion (the solid line). Then there are two other groups of people who are very high in extraversion (the dotted line), and those who are very low in extraversion (the dashed line). For the time being, ignore the "mean" extraversion (the solid line) group and just look at the two extreme groups. Here, we see there is a pronounced difference with the way that extraversion influences those two groups of people. Compared to introverts, those who are extraverted are much more likely to judge a person free and morally responsible in the high affect case compared to the low affect case. That, in essence, is the nature of the

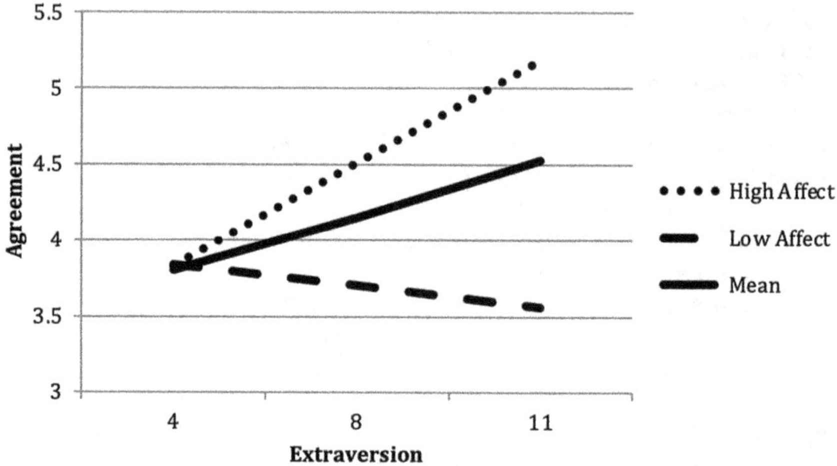

Fig. 2.4 Personality moderating judgments of free will and moral responsibility between high and low affect cases

interaction. However, now imagine what the graph would look like if you averaged to two extreme groups. If you imagined that the average would look like the solid black line, then you are right. If you average all responses, you will find a very small overall effect even though it is a fiction produced by averaging differences in some identifiable groups (e.g., those high in extraversion) who are influenced by affect and some that were not (e.g., those low in extraversion).

These results support our prediction concerning why the effect of affect on people's free will and moral responsibility judgments is typically small. Only some people are influenced by affect whereas others are not. Those who are high in the personality trait extraversion are relatively strongly influenced by the affective content of the scenarios whereas introverts are not influenced by the affect (if anything, there is a non-statistically significant numerical shift in the opposite direction). When these groups are combined, introverts mute the effect of affect. The result is an overall small effect that is sometimes difficult to detect with overall mean scores. As such, any model that does not take into account individual differences risks modeling a fictitious "average" person. In this case, "the average" does not seem to accurately represent any of the actual subgroups of respondents, as many people tend to have more (or less) polarized sets of intuitions as compared to the average. This is like what you might find by averaging people's responses on surveys about their political orientation. The "average" person may be politically moderate. However, most people exist closer to political extremes. Only focusing on the "average" thus distorts or neglects people closer to the extremes, and so offers a distorted or otherwise inaccurate description of response patterns.

Conclusion

This chapter has reviewed a diverse body of scientific data indicating that personality predicts intuitions about freedom and moral responsibility. The relation between extraversion and compatibilist judgments is found in different labs, using different materials, with different response options, measured cross-culturally with diverse samples of people from various backgrounds who speak different languages. Across all studies, research consistently finds that personality predicts free will and moral responsibility intuitions about the compatibility question, manipulation, and some fated actions.

These results are not only empirically interesting, they also have some philosophical bite. One theoretical assumption is that intuitions are invariant (i.e., they do not change much from place to place, time to time, or person to person) (Knobe & Doris, 2010). There are at least three kinds of variance. *Inter-cultural* variance occurs when members of different cultures (e.g., Easterners and Westerners) have different intuitions. *Intra-cultural* variance occurs when individuals of the same culture have different intuitions (e.g., extraverts in America have different intuitions than introverts in America). *Inter-temporal* variance occurs when the same individual has different intuitions at Time 1 and Time 2. These three types of variances are logically independent. They could all be true, they could all be false, or some may be true while some may be false. Do the studies presented in this chapter inform what types of variability there is with free will and moral responsibility intuitions? To more precisely answer this question, we'll concentrate on variability associated with the compatibility question.

Take inter-cultural variability. Quite a lot has been made of inter-cultural variability in people's intuitions about a variety of subjects. However, it appears that there are some culturally universal intuitions about freedom and moral responsibility. Sarkissian et al. (2010) found that there was not much inter-cultural variability when the compatibility question was framed abstractly. Most people across cultures had incompatibilist friendly intuitions. Schulz et al. (2011) found that intuitions about concretely described individuals appeared to be compatibilist in a German-speaking European sample. These two studies suggest that the abstract/concrete difference may exist in different cultures. Consequently, these data suggest cross-cultural stability for at least some prominent effects in the experimental philosophy of free will.[15] Admittedly, there is a relatively small amount of data about inter-culture reliability of personality's relation to free will judgments. But if future work resembles past work, we should continue to see these relations cross-culturally.

While there appears to be some inter-cultural stability about the compatibility question, the same cannot be said about intra-cultural stability. In some instances, there is substantial divergence in intuitions of people within the same culture. For example, the relation of extraversion with compatibilist intuitions in Spanish, English, and German-speaking samples suggests that there is consistent intra-cultural variance in intuitions about the compatibility question. The inter-cultural stability and intra-cultural variability associated with personality is to be expected. Recent studies suggest that the members of the Big Five personality traits are present across almost all cultures and have roughly the same distribution, especially in Western cultures (Schmitt et al., 2007). Because extraversion is associated with many compatibilist intuitions and extraversion is present in many cultures, we should expect that (a) intuitions about the compatibility question are stable cross-culturally and (b) that within the culture, personality would predict intuitions about the compatibility question.

The studies in this chapter also reveal at least some short-term inter-temporal stability. When presented with both High and Low Affect Determinism scenarios, people tend to have incompatibilist intuitions about both scenarios and the short-term stability of judgments between these scenarios is high ($r = .75$) (A. Feltz & Cokely, 2019). The relation to global personality traits bolsters the inter-temporal stability of intuitions about freedom and moral responsibility. While there is some evidence of personality traits change somewhat over the course of life, especially for younger people, after age 30 there is relatively little change in personality traits relative to their own cohort (McCrae, 2002). Differences in

[15] Of course, there may be some domains where there are systematically different intuitions between cultures. For example, perhaps there are different moral intuitions (Sommers, 2012) or intuitions about reference (Machery et al., 2004).

personality also tend to be relatively stable even when they change (e.g., older adults will tend to be moderately more conscientious than younger adults; however, someone who is moderately conscientious will stay that way most of their life relative to people of the same age (Ashton, 2013)). Moreover, many of global or general personality traits (e.g., extraversion) are strongly heritable (Bouchard & Loehlin, 2001; Jang, Livesley, Angleitner, Riemann, & Vernon, 2002; Luo, Kranzler, Zuo, Wang, & Gelernter, 2007; Spinath & Johnson, 2011; Wilt & Revelle, 2009). Some of the estimates suggest that the heritability of personality traits (i.e., the amount of variance in a population that is attributable to genes) is about 50% (Bouchard & McGue, 2003). This means that not only are similar intuitions about the compatibility question likely to persist over one's life, one may be quite likely to pass on the tendencies to have those intuitions to the next generation. In the light of evidence that even experts show similar patterns of bias (see Chap. 5), disagreement about the compatibility question seems very likely to be persistent and trans-generational across cultures.

Of course, the personality traits associated with free will intuitions could be culturally variable (e.g., personality and culture could interact). For example, East Asians tend to be less extraverted than people in other parts of the world (Schmitt et al., 2007). This would suggest that there should be fewer individuals in that part of the world who respond that one could be free and morally responsible for concretely described compatibility questions. However, the effect of culture on the relation between personality and compatibilist judgments is likely to be small since there is an overall modest effect of culture on extraversion (about 3% of the variance) (Schmitt et al., 2007). Future research is needed to determine if, how, or where personality and culture interact to generate intuitions about the compatibility question.

Taken altogether, the evidence and statistical models we have reviewed strongly suggest to us that intuitions about free will and moral responsibility are temporally stable, pervasive (inter-culturally stable), and likely to be very similar around the world (intra-cultural variability). For some people, judgments about freedom and moral responsibility are robust to a number of different threats to freedom and moral responsibility such as determinism, moment-by-moment manipulation, fate, or being compelled to act by a neurological condition (De Brigard, Mandelbaum, & Ripley, 2009). However, others are sensitive to these threats to free will. Consequently, there is a spectrum of intuitions about freedom and moral responsibility that are a function of a variety of factors. Nevertheless, many of these intuitions are associated with heritable personality traits that are known to be largely stable across cultures and lifespan.

Appendix

"Spanish Description of Determinism

Por favor lea el siguiente pasaje cuidadosamente y responda a las preguntas que lo siguen.

La mayoría de psicólogos están convencidos que eventualmente vamos a descubrir exactamente como todas nuestras decisiones y acciones están enteramente causadas. Por instancia, ellos piensan que siempre que estamos intentando decidir qué hacer, la decisión que terminamos por hacer es completamente causada por los pensamientos, deseos y planes específicos ocurriendo en nuestras mentes. Los psicólogos también están convencidos que estos pensamientos, deseos y planes son completamente causados por nuestra situación corriente y los eventos anteriores en nuestras vidas y que estos eventos anteriores también son completamente causados por eventos más anteriores, eventualmente yendo hasta eventos que ocurrieron antes de que naciéramos.

Por lo tanto, una vez que eventos anteriores específicos han ocurrido en la vida de una persona, estos eventos definitivamente van a causar que ocurran eventos posteriores específicos.

Spanish High Affect

Por ejemplo, un día una persona llamada Juan decide matar a su esposa para poder casarse con su amante, y lo hace. Una vez que los pensamientos, deseos y planes específicos ocurren en la mente de Juan, ellos definitivamente causan su decisión de matar a su esposa.

Suponga que los psicólogos están acertados que los eventos que ocurrieron anteriormente en la vida de Juan causaron su decisión de matar a su esposa.

Por favor califique a que grado está usted de acuerdo con las siguientes exposiciones.

1. Juan matando a su esposa dependió de él. (1—totalmente en desacuerdo 7—totalmente de acuerdo)
2. Juan mató a su esposa de su propia y libre voluntad.
3. Juan es moralmente responsable por matar a su esposa.

4. ¿Si los psicólogos están correctos, es exacto decir que si el universo fue recreado Juan haría la misma cosa? (Si, No)

Spanish Low Affect

Por ejemplo, un día una persona llamada Juan decide hacer trampa en sus impuestos, y lo hace. Una vez que los pensamientos, deseos y planes específicos ocurren en la mente de Juan, ellos definitivamente causan su decisión de hacer trampa en sus impuestos.

Suponga que los psicólogos están acertados que los eventos que ocurrieron anteriormente en la vida de Juan causaron su decisión de hacer trampa en sus impuestos. Por favor califica a que grado usted está en acuerdo con las exposiciones que siguen.

1. Juan haciendo trampa en sus impuestos dependió de él. (1—totalmente en desacuerdo 7—totalmente de acuerdo)
2. Juan hizo trampa en sus impuestos de su propia y libre voluntad.
3. Juan es moralmente responsable por hacer trampa en sus impuestos.
4. ¿Si los psicólogos están correctos, es exacto decir que si el universo fue recreado Juan haría la misma cosa? (Si, No)"

Book

"*Spanish Fatalism Description*

Por favor lea el siguiente pasaje cuidadosamente y responda a las preguntas que lo siguen.Hay un libro especial que tiene todas nuestras decisiones y acciones verdaderamente escritas en su contenido. Por instancia, siempre que estamos intentando decidir qué hacer, la decisión que terminamos por hacer es completamente y verdaderamente escrita en este libro. Este libro especial tiene verdaderamente escrito en el estos acontecimientos siglos antes de que el evento se lleve a cabo. Por lo tanto, si el libro tiene un evento escrito en él, el evento definitivamente va ocurrir.

Spanish Fatalism High Affect

Por ejemplo, un día una persona llamada Juan decide matar a su esposa para poder casarse con su amante, y lo hace. Una vez que el evento específico está verdaderamente escrito en el libro, es imposible para Juan no matar a su esposa.

Suponga que el contenido del libro lo hico imposible para Juan no matar a su esposa. Por favor califique a que grado está usted de acuerdo con las siguientes exposiciones.

1. Juan matando a su esposa dependió de él. (1—totalmente en desacuerdo 7—totalmente de acuerdo)
2. Juan mató a su esposa de su propia y libre voluntad.
3. Juan es moralmente responsable por matar a su esposa.
4. ¿Si el universo fue recreado con el libro especial teniendo las mismas verdaderas oraciones Juan haría la misma cosa? (Si, No)

Spanish Fatalism Low Affect

Por ejemplo, un día una persona llamada Juan decide hacer trampa en sus impuestos, y lo hace. Una vez que el evento especifico está verdaderamente escrito en el libro, es imposible para Juan no hacer trampa en sus impuestos. Suponga que el contenido del libro lo hico imposible para Juan no hacer trampa en sus impuestos. Por favor califique a que grado está usted de acuerdo con las siguientes exposiciones.

1. Juan haciendo trampa en sus impuestos dependió de él. (1—totalmente en desacuerdo 7—totalmente de acuerdo)
2. Juan hizo trampa en sus impuestos de su propia y libre voluntad.
3. Juan es moralmente responsable por hacer trampa en sus impuestos.
4. ¿Si el universo fue recreado con el libro especial teniendo las mismas verdaderas oraciones Juan haría la misma cosa? (Si, No)

Spanish Comprehension Questions (Correct answers in bold)

1. A los perros les gusta ladran, a los gatos les gusta dormir, y a los caballos les gusta correr. ¿Cual es la palabra incorrecta en esta frase?

   (a) **Ladran**
   (b) Gusta
   (c) Gatos
   (d) Les

2. Las puertas normalmente están hechas de sopa.

   (a) Cierto
   (b) **Falso**

3. Es costumbre saludarle a personas conocidas.

   (a) **Cierto**
   (b) Falso

4. Si juego al ajedrez uso…
    (a) **Un tablero**
    (b) Una naranja
    (c) Un pájaro
    (d) Un microondas
5. No puede uno viajar en
    (a) Bote
    (b) Avión
    (c) Tren
    (d) **Águila"**

CHAPTER 3

Intentions and Side Effects

"Suppose the vice-president of a company went to the chairman of the board and said, 'We are thinking of starting a new program. It will help us increase profits for this year's balance sheet, but in ten years it will start to harm the environment.' The chairman answered, 'I don't care at all about harming the environment. I just want to make as much profit for this year's balance sheet as I can. Let's start the new program.' They started the new program. Sure enough, the environment started to be harmed. Did the chairman intentionally harm the environment?"

Judgments of intentions and intentionality are fundamental elements of social cognition (Malle, Moses, & Baldwin, 2001). To see this, just imagine the different reactions you would have if somebody accidentally versus intentionally stepped on your foot in an elevator. You'd likely have a stronger, less forgiving attitude toward the intentional foot stepper. As you can see, judgments of intentionality are pervasive and important in everyday life. Intentionality judgments may also profoundly impact people's lives. For example, in court cases, if somebody is thought to have intentionally brought about harm, greater amounts of punishment or remuneration are often warranted. Or, if you found out your mother intentionally killed somebody you would probably feel differently than you would if you learned she accidentally killed somebody. As such, judgments of intentionality can play important roles in how we relate, react, and interact with others.

Much like the free will debate, many theorists hold that there is, or should be, a tight connection between the philosophical and empirical

A. Feltz, E. T. Cokely, *Diversity and Disagreement*,
https://doi.org/10.1007/978-3-031-61935-9_3

investigation of intentions and intentionality. Some philosophers are explicit that they take folk intuitions seriously (Adams, 1986; McCann, 1998, 2005; Mele, 1992). As Mele writes, "a philosophical analysis of intentional action that is wholly unconstrained by that [folk] concept runs the risk of having nothing more than a philosophical fiction as its subject matter" (A. Mele, 2001, p. 27). Mele continues that "what it is to do something intentionally.....will be anchored in common-sense judgments about particular hypothetical or actual actions" (2001, p. 27).

Partially because of this tight connection between everyday intuitions and theorizing about intentional action, a sizable empirical literature has sprung up in the past few decades. This literature about intentional action has largely dealt with what we will call the Knobe effect or side-effect effect (Knobe, 2003a).[1] The Knobe effect is typified by people judging that bad side effects (e.g., harming the environment) are brought about intentionally whereas good side effects (e.g., helping the environment) are not.

This chapter revolves around personality predicting the Knobe effect. However, some other theoretical side effects of these data are worth mentioning. We therefore start with a review of some of the major explanations of Knobe effect. We argue, similar to the arguments in Chap. 2, that no single account of the Knobe effect (or intentional action, for that matter) is likely to accurately capture all or most intuitions about intentional action. Instead, the Knobe effect appears to multiply determined by a number of factors including biases, concepts, personality, and judgment environments. As such, any single parsimonious account of the Knobe effect is not likely to succeed. The data about the Knobe effect again highlight the dangers of treating "the folk" as a monolithic entity. Intuitions are importantly diverse and predictable, and this diversity is philosophically and practically important.

The Intentional Action Side-Effect Effect

Defining "side effect" can be somewhat tricky. We will use the following definition of a side effect: "*X* is a side-effect action performed by an agent *S* if and only if *S* successfully seeks to perform an action *A*, *E* is an effect of

[1] The "Knobe effect" has also sometimes referred to a family of effects where moral goodness or badness influences other types of (mainly non-moral) judgments such as causation, doing, allowing, knowing, to name just a few (see, for example, Knobe (2010b)).

his so doing, *X* is his bringing about *E*, and *X* has the following properties: *S* is not at the relevant time seeking to *X* either as an end or as a means to an end, and *X* is not in fact a means to an end that *S* is seeking at the relevant time" (Cushman & Mele, 2008, p. 179). Judgments of side effects have been thought to be important because they feature events that are foreseen but not intended (McCann, 2005; Nadelhoffer, 2006a). For example, one prominent view of intentional action—the Simple View—holds that side effects can never be brought about intentionally (Adams, 1986; McCann, 1998). The Simple View states that an action is intentionally brought about only if that action is intended. Side effects are by definition consequences of actions that are not intended, even if they are foreseen. For these reasons, on the Simple View, side effects of actions cannot be brought about intentionally. However, many people judge some side effects to be intentional (Knobe, 2003a, 2003b, 2004, 2006, 2010a, 2010b; Mele & Cushman, 2006; Nadelhoffer, 2006a, 2006b), which has been argued to count as a strike against the Simple View of intentional action (Nadelhoffer, 2006b).

Knobe (2003a) provides evidence that intentional action intuitions about side effects are responsive to the "goodness" or "badness" of the side effect. That is, when participants are presented with a good side effect, they tend not to judge that side effect is brought about intentionally. However, when people are presented with a bad side effect, they tend to judge that bad side effect as being brought about intentionally. This asymmetry is the Knobe effect or the side-effect effect. To illustrate, consider the following pair of cases:

> *Harmful/Helpful Chairman*
>
> The vice-president of a company went to the chairman of the board and said, "We are thinking of starting a new program. It will help us increase profits for this year's balance sheet, but in ten years it will start to *(harm/help)* the environment." The chairman answered, "I don't care at all about *(harming/helping)* the environment. I just want to make as much profit for this year's balance sheet as I can. Let's start the new program." They started the new program. Sure enough, the environment started to be *(harmed/helped)*. (Knobe, 2003a)

Theoretically, the main difference between Harmful and Helpful Chairman is the moral valence of the side effect. The helpful side effect is good and the harmful side effect is bad. When participants were asked to respond to the following statement "the chairman intentionally harmed/ helped the environment," their responses were qualitatively different (help is judged unintentional, harm is judged intentional). This effect has been replicated several times across cultures (Knobe & Burra, 2006), ages (Leslie, Knobe, & Cohen, 2006), and scenarios (Cushman & Mele, 2008; Knobe, 2003b, 2004, 2006; Mele & Cushman, 2006; Nichols & Ulatowski, 2007).

What could account for these differences between the helpful and harmful Chairman Cases Two prominent general types of explanations have been offered—core concept and non-concept explanations. Core concept explanations attempt to explain the difference as a function of the proper application of our concept of intentional action. That is, our concept of intentional action is somehow sensitive to the harmfulness or helpfulness of the action, and that helpfulness or harmfulness of the action is an appropriate element used in classifying whether an action is intentional. Non-concept explanations do not hold that our core concept of intentional action is appropriately sensitive to the moral valence of the action. Rather, the difference between the helpful and harmful Chairman Cases is the result of something other than the correct application of the concept such as making a mistake. As such, responses to either the helpful or harmful Chairman Cases do not necessarily reflect the correct application of a concept of intentional action and require additional mechanisms to account for the difference. We now turn to illustrative examples of each type of explanation.

Correctly Applied Concept Explanations

Knobe (2006) attempts to explain the asymmetry by different systems that are activated for the good and bad side effects. In the first process, one identifies the side effect as being either good or bad. For example, in this process, people identify the harmful chairman as bringing about the harmful side effect and identify the Helpful chairman as bringing about the helpful side effect. After this determination, one uses one's concept of intentional action. If properties sufficient for intentionality judgments are found (e.g., foresight for bad side effects, desire for good side effects), then one judges that those good or bad side effects are intentionally brought about. Given this account, we can explain the asymmetry. Moreover, on this account the core concept of intentional action plays an important, and appropriate, role. It is part of the very concept of intentional action that it looks for moral features of the actions (and consequences of actions) that are to be judged intentionally. As such, this account gives central role to one's concept of intentional action. As Knobe himself notes, "moral considerations are playing a helpful role in people's underlying competence itself" (Knobe, 2006, p. 226).

Phelan and Sarkissian (2008) object to Knobe's (2006) explanation. In one study, they provided participants slightly modified, but structurally similar, chairman scenarios (Knobe & Mendlow, 2004). These scenarios describe the president of a corporation who intends to increase sales in Massachusetts and foresees, but doesn't care, that it will decrease sales in New Jersey. On a natural reading, decreasing sales in New Jersey is a side effect of increasing sales in Massachusetts. Most people judged that the president lowered sales in New Jersey intentionally. On Knobe's original view, the judgment of intentionality should mean that the participants view the side effect as bad. But they don't. Most people did not judge lowering sales tax in New Jersey as bad. The lack of judging the side effect as bad and the presence of the asymmetry puts pressure on Knobe's (2006) account since the first system would not provide different verdicts for the two side effects. Hence, at least in some cases, the two-system account cannot completely explain the asymmetry typical of the Knobe effect.

Some mental states of the actor also influence ascriptions of intentionality to side effects. One mental state that reduces intentionality judgments of bad side effects is regretfully bringing about that side effect (Phelan & Sarkissian, 2008). Sverdlik (2004) set up an experiment where a person foresees that mowing the lawn will wake up his neighbors, but he does it anyway. In one condition, he is described as regretfully waking up his neighbors. In the other condition, any mention of regret is omitted. Participants were randomly assigned to one of the conditions. Those in the regret condition had significantly lower intentionality ratings for waking up the neighbors than those in the no-regret condition. Much like the results of the Phelan and Sarkissian study, these results put pressure on the two-process account since in both cases the side effect was bad. Hence, the badness of the side effect and activation of corresponding processes again do not completely explain intentionality judgments of side effects.

These types of results have forced a revision in Knobe's (2006) view. Pettit and Knobe (2009) have offered the following revised concept-based account of the Knobe effect. In general, many people are sensitive to whether the person being evaluated acts with an appropriate or inappropriate attitude. Something similar may happen when people make intentionality judgments. In particular, we can think of the attitude that the agent has as a spectrum from having a favorable to unfavorable attitude toward a side effect. The agent's attitude can fall anywhere on that spectrum. The moral judgments that one makes about the side effect and the agent's attitude to that side effect set the "default" point on the spectrum to which intentionality judgments are compared and made. If the agent's attitude falls on the favorable side of the default, the action is judged intentional. If the agent's attitude falls on the unfavorable side of the default, the action is judged as unintentional. But the important thing is that the default position can change in part depending on the moral judgment one makes about what attitudes one should have toward an event. According to Pettit and Knobe (2009), this can explain the Knobe effect even though the chairman has exactly the same attitude (i.e., not caring) in both Helpful and Harmful Chairman. In Helpful Chairman, the chairman lacks a favorable attitude toward something that he normally should have a favorable attitude toward, so his attitude falls on the unfavorable side of the scale and is judged unintentional. In Harmful Chairman, the chairman lacks a negative attitude that he normally should have toward harming the environment, so his attitude falls on the favorable side of the default and is judged intentional. Similar reasoning can be applied to explain the Sverdlik and the Phelan and Sarkissian studies.

Nichols and Ulatowski (2007) have an alternative view of the side-effect effect. Their view centers on "interpretative diversity." As you will see, this view significantly shaped our own views about the Knobe effect (and other effects reported in experimental philosophy). This account holds that people interpret the word "intentionally" differently. Some people have an interpretation that focuses on foresight while others have an interpretation that focuses on desire. Others oscillate between these two interpretations, thus accounting for the asymmetric pattern of judgments typical of the side-effect effect. To test their view, participants responded to both the Harmful Chairman and Helpful Chairman cases. Their results again confirmed the existence of a judgment asymmetry, and they, like previous studies, found that many people *did not* display the judgment asymmetry. However, and important for Nichols and Ulatowski's view, in their studies roughly a third of all participants responded "no" to both Harmful and Helpful Chairman, a third responded "yes" to Harmful and Helpful Chairman, and a third responded "yes" to Harmful Chairman and "no" to Helpful Chairman. Participants were allowed to explain their answers. Two major explanations emerged. One explanation is that the chairman knew that the side effect was going to come about. These people tended to think that the chairman brought about the side effects intentionally. The other explanation involved the desire of the chairman. Since the chairman did not desire to bring about the side effect, participants using this explanation tended to think that the side effect was not brought about intentionally. Hence, Nichols and Ulatowski concluded that "considerations of outcome may influence which interpretation the term is given" (2007, p. 361). In terms of these explanations, it seemed plausible to think that people employ a knowledge-based and a motive-based concept of intentional action.

Cushman and Mele (2008; Mele & Cushman, 2006) provide additional evidence and extend Nichols and Ulatowski's work. In their studies, in addition to the two Chairman Cases, participants were given a variety of scenarios focusing on two major differences. In one set of scenarios, the person is described to encourage the interpretation that the person believes that side effect will come about but does not desire that side effect. In the other set of scenarios, the person is described to encourage an interpretation that the person desires the side effect come about but does not believe that it will. Like Nichols and Ulatowski, Cushman and Mele found three general patterns of responses—some answer "yes" to both Harmful and Helpful Chairman, some answer "no" to both, and some answer "yes" to Harmful Chairman but "no" to Helpful Chairman. Because their vignettes systematically varied belief and desires, they discovered that nearly all people think that if a person desires a side effect to come about (and the side effect does come about), then they did it intentionally. However, there was diversity concerning whether belief was enough to judge that a side effect is brought about intentionally, with some people thinking that belief is enough whereas others thought that belief was not enough to judge that side effect as being brought about intentionally. For these reasons, Cushman and Mele think that there are at least two concepts of intentional action—one where belief is enough to bring about a side effect intentionally and one where desire is required to bring about a side effect intentionally. Cushman and Mele also allow for the possibility of a third concept where normally desire is required for an action being judged intentionally except for actions that are morally bad. In those morally bad instances, knowledge may be enough to judge the action intentional. This third concept can thereby explain the judgment asymmetry typical of the side-effect effect.

Sripada (2010) attempts to explain the Knobe effect with what he calls the Deep Self Model. The Deep Self is the stable set of psychological characteristics of a person. Contrast the Deep Self with the Acting Self. The Acting Self is the self that is the proximal cause for an action. Often the Deep and Acting Self correspond to one another, for example, when a vegetarian does not eat meat on a particular occasion. But sometimes the Deep and Acting self are discordant—for example, when the momentary desire to eat an expertly prepared piece of meat just overwhelms the vegetarian and she eats the meat. According to Sripada, people's concept of intentional action involves the Concordance Criterion: when the event that one's action brings about lines up with one's deep self, then we are likely to judge the outcome intentional. As such, Sripada argues that normative considerations (goodness or badness of the side effect) are irrelevant to the Knobe asymmetry. The chairman in Harmful Chairman is more likely to be thought to have a deep and enduring disdain for the environment than the chairman in Helpful Chairman. Sripada presents evidence that the "Deep Self" accounts for the asymmetry because after controlling for Deep Self features, the normative features (goodness/badness, blame/praise) are not related to the asymmetry. Hence, the asymmetry is explained without reference to normative factors at all but rather by one's concept of intention action.

But the Deep Self model is not completely satisfactory (see also Rose, Livengood, Sytsma, and Machery (2012)). Jason Shepard (2011) has argued that while the Deep Self model can explain some of the variance associated with the asymmetry, it cannot explain it all. In fact, non-normative factors may still play a role in people's concept of intentional action. Shepard offers a number of examples where deep self-attitudes did not have a significant impact on intentionality judgments in accordance with the Deep Self model. To take one example, Shepard gave participants a scenario indicating that the chairman had a deep and abiding commitment to the environment. In such a case, the Deep Self model should predict that the chairman intentionally brought about the help to the environment because he has a Deep Self in concordance with the event that is brought about by his intended action. However, Shepard's experimental results did not bear out this prediction. In fact, the results in the caring chairman case were not different from Helpful Chairman, calling into question the Deep Self model.

Not Correctly Applied Concept-Based Explanations

Unlike concept-based explanations that take the Knobe effect to reflect a central feature of competent application of the concept of intentional action, non-concept explanations try to explain the Knobe effect without reference to our concept of intentional action. Non-concept explanations can proceed in a number of ways. For example, as we will see, non-concept explanations can hold that people are somehow illegitimately biased when making intentional action ascriptions. Or perhaps people really mean something else when they use the word "intentional" in the contexts that generate the Knobe effect. If they mean something else with the word "intentional," then again, they do not apply their core concept of intentional action in the Chairman Cases. Rather, people use some other concept and use the label "intentional" as a way to talk about that concept given the restrictions in the experiment (e.g., questions asking about intentionality). In these ways, people are not correctly applying their concept of intentional action.

Several non-concept accounts have been offered to explain the side-effect effect of which we will review just a few. One is offered by Edouard Machery (2008). On Machery's account, the asymmetry can be explained by the notion of trade-offs. In the Harmful chairman case, the chairman can be seen as trading off a bad thing (i.e., harming the environment) for a good thing (i.e., increasing the bottom line). Because there is a bad thing that occurs with a good thing, many people may view that kind of exchange as being intentionally brought about. However, in the Helpful chairman case, there is no bad thing being traded for a good thing. Rather, the side effect is helpful and need not be traded for. So, in that case, many people may be inclined to think that the helpful side effect is not intentionally brought about. Hence, the trade-off hypothesis can explain the side-effect effect with having to reference to core concept of intentional action (however, see Machery (2008) for potential problems with this account).

On an alternative account, Adams and Steadman (2004a, 2004b) focus on potential conversational implicature that may take place in the Chairman Cases (Grice, 1975). Conversational implicature can occur when, given the appropriate context, one uses a word or a phrase to express something other than what is literally meant by that word or phrase. Take, for example, a basketball game. A player may take a horrible shot and a fan may say "nice shot" to express displeasure with the shot. The fan uses the expression "nice shot" to mean something else (in this case, conversationally implying "bad shot"). Adams and Steadman argue that something similar may happen in the Chairman Cases. Participants may judge that the Harmful Chairman is blameworthy for bringing about the bad side effect. Given the response options in the experiment, participants can only express this judgment of blame by indicating that the chairman brought the bad side effect about intentionally. If participants did not indicate the harm was brought about intentionally, then that would conversationally imply that the chairman is not blameworthy. For the Helpful chairman, participants do not want to conversationally imply that the chairman is praiseworthy for not caring about helping the environment. Consequently, they judge helping the environment is not intentionally brought about. In these ways, conversational implicature, and not a core concept, can explain the side-effect effect (see Malle (2006) and Knobe (2003b) for responses).

Nadelhoffer (2004) thinks that the side-effect effect is best explained by biasing. In this case, the bias is the result of differential affective reactions that people tend to have toward the Helpful and Harmful Chairman. In the Harmful Chairman case, there is negative affect that is generated by the chairman not caring about harming the environment. Participants thereby perceive the Harmful Chairman negatively and this negative impression (caused by the bias) results in participants tending to judge that the harm was brought about intentionally. Along a similar line, people are likely to have a negative affective reaction toward the Helpful Chairman because the Helpful Chairman does not care about something that he should (i.e., helping the environment). Because of the bias, people do not want to praise the Helpful chairman for bringing about the help but they do want to blame the chairman for bringing about the bad side effect. When Nadelhoffer measured praise and blame ratings for the Helpful and Harmful chairman (respectively), praise ratings were much lower than blame ratings. Hence, these data are consistent with an affective biasing account that could explain the side-effect effect (see Nadelhoffer (2004, 2006a) for direct evidence for the affective biasing account).

Similar to Nadelhoffer (2004), Malle and Nelson (2003) propose an account where negative affect generated in the Harmful chairman case plays a role in judging that side effect intentional. They key on the tendency for people, when they are fighting, to judge the person who they are fighting with harshly. These judgments, including intentionality judgments, are biased by the negative affect that is generated during the fight. For example, if Jean and Robin are fighting and Jean bumps into a plate and breaks it, there is a greater tendency for Robin to think that Jean broke the plate intentionally.

Something similar may be happening in the Chairman Cases. Malle (2006) argues that as the Chairman Cases are described, the Harmful Chairman does not care at all about the bad side effect he brings about. Given that bit of information, participants may think that they are supposed to use that information in the judgment about the chairman. After all, it is natural for participants to think that critical information included in a scenario is there for a reason. If participants think that they are supposed to use that information, they may use that evaluative information rather than their core concept of intentional action when making judgments about the Chairman. That would mean that people have the tendency to judge the Harmful Chairman as bringing about the side effect intentionally. However, in the Helpful Chairman case, participants may be less likely to think that they are supposed to use the information about helping the environment, and thereby judge that the Helpful Chairman did not bring about the side effect intentionally.

Personality Predicts Philosophical Disagreement in Intentional Action

At this point, we have documented two different general types of explanation for the side-effect effect (and admittedly not an exhaustive recounting of all possible explanations). We want to emphasize that these studies (and others) had substantial dissenting minorities. These dissenting minorities call for an explanation of why they respond differently from the majority responses. To begin to help offer a potential explanation, we will detail some of the experimental work concerning the relation of personality traits to the Knobe effect in section "The Knobe Effect and Extraversion." In sections "The Knobe Effect, Extraversion, and Theoretical Accounts" and "Conclusion" we will then discuss why the empirical data could be important for theoretical accounts of the Knobe effect discussed in the previous sections.

The Knobe Effect and Extraversion

As already detailed in previous chapters, many personality traits are related to differences in cognition. These differences include one's motivation, judgment tendencies, detection of cues in the environment, the perceived importance of those cues, among others (Funder, 1991, 1995; McCrae & Costa, 1990). Our focus here is again on the global personality trait extraversion. Extraverts are socially minded individuals with relatively less emotional regulation who are motivated to engage in social activities and cognition. It stands to reason, then, that when a person engages in a harmful, socially undesirable action, extraverts would be more likely than introverts to be sensitive to those undesirable events and more motivated to express their feelings. As such, the Knobe effect is likely to be positively related to extraversion since harming the environment is socially undesirable whereas helping the environment is not.

The first study we review focuses on the classic Knobe effect Chairman Cases (A. Feltz & Cokely, 2009). A similar procedure used in the free will studies was used to establish the relation of extraversion with the Knobe effect. Participants were given the classic chairman scenarios and a general measure of global personality traits including extraversion (the Ten Item Personality Inventory). It was predicted that extraversion would be related to the side-effect effect. However, extraversion is likely not the only factor that contributes to the side-effect effect. Other individual differences are likely related as well, such as expertise or cognitive impulsivity. Because these other factors are known to influence judgments in other domains they may also contribute to the judgments typical of the side-effect effect (Cokely, Galesic, Schulz, Ghazal, & Garcia-Retamero, 2012; Ericsson, Prietula, & Cokely, 2007; Frederick, 2005; Stanovich, 1999). For example, those who have a reflective cognitive style are more likely to wait some time to receive a larger reward and cognitively reflective individuals may be less likely to be influenced by framing effects in gambles than less cognitively reflective individuals (Frederick, 2005). Individual differences in attentional control are related to differences in strategies such as double checking and are associated with longer deliberation about problems or tasks (Cokely & Kelley, 2009; E. T. Cokely, Kelley, & Gilchrist, 2006). Any of these factors could influence some people to think more carefully about the cases, and that could cause different patterns of judgments about side effects (e.g., being less biased).

Accordingly, cognitive impulsivity was measured and statistically controlled for in the first study we will review. Cognitive impulsivity refers to some individuals' tendency to rely on intuitive or gut reactions versus those who tend to rely on more deliberate and effortful processing. We used the Cognitive Reflection Task to measure cognitive impulsivity (Frederick, 2005). We also controlled for other potential individual differences by measuring self-control (Tangney, Baumeister, & Boone, 2004); working memory capacity (as measured by the working memory operation-span measure (OSPAN)), (Turner & Engle, 1989); and self-reported scholastic aptitude test (SAT) scores which are known to correlate with general intelligence (Frederick, 2005). Consistent with what we have done in previous chapters, we used the Ten Item Personality Inventory to measure the Big Five personality traits (Gosling et al., 2003). Finally, participants responded to both the Harmful and Helpful Chairman Cases, counterbalanced for order on 7-point scale (from strongly disagree to strongly agree) (Mele & Cushman, 2006).

A mixed model analysis of variance (ANOVA) with order (Help first, Help second) as a between-subjects variable and side effect (Harm, Help) as a repeated measure revealed a large difference in side-effect intentionality judgments (i.e., asymmetry). Harmful side effects were judged intentional ($M = 5.0$, $SD = 1.9$) whereas helpful side effects were judged unintentional ($M = 2.1$, $SD = 1.5$).[2] To provide evidence that extraversion had unique predictive power when controlling for other individual differences, a stepwise multiple linear regression with the side-effect asymmetry as the dependent variable and personality, brief self-control, OSPAN, SAT, CRT, and sex as independent variables was conducted. The analysis revealed that only extraversion was reliably related to the judgment asymmetry.[3]

To further illustrate the observed relations between the Knobe effect and extraversion, were created extraversion quartiles. To create the extraversion quartiles, we looked at the overall distribution of extraversion scores for the participants. Then we identified the top 25% and the bottom 25% of the extraversion scores. We used those extraversion quartiles as independent variables in an ANOVA. This analysis revealed a large overall interaction of extraversion with judgments about the Chairman Cases. In short, those who were extraverted displayed a large difference in judgments typical of the side-effect effect whereas those who were introverted had a much more muted asymmetry (see Fig. 3.1).

One of the take-home messages of this study was that extraversion was related to the side-effect effect. Moreover, even after statistically controlling for several individual differences, extraversion continued to predict

[2] $F(1, 93) = 148.24$, $p = .001$, $\eta_p^2 = .61$.

[3] $\beta = .29$, $t = 2.46$, $p = .02$, $R^2 = .08$. All other individual differences were unreliable (Fs < 1).

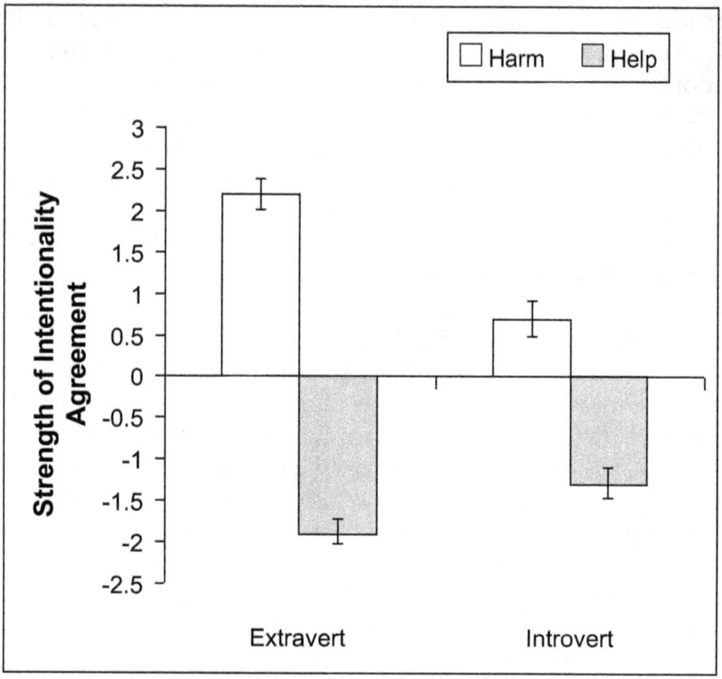

Fig. 3.1 Introverts and Extraverts (bottom versus top quartile) by side-effect intentionality ratings (harm, help). Positive numbers indicate agreement, negative numbers disagreement, and error bars represent the standard error of the mean

the judgment asymmetry. This provides some evidence that extraversion is perhaps the most important general individual difference that predicts the side-effect effect (at least in some paradigmatic cases).

Replications

If the relation of extraversion with the Knobe effect is real, then it should be found in subsequent studies that attempt to estimate that relation. We report three new studies to replicate the effect of extraversion with the Knobe effect.

In the first study, we used a probabilistically representative national sample of the United States recruited from the company Knowledge Networks (N = 295), which conducted a survey with a diverse range of people living in the United States that was essentially proportional to the actual diversity of people living in the United States (i.e., probabilistically representative). The mean age was 46, $SD = 16.15$ and 50% identified as female. Just as in the previous study, participants completed the two Chairman Cases presented in the harm-help order to help reduce the known order effect. Participants responded on a 7-point Likert scale (1 = strongly disagree, 7 = strongly agree). Participants also completed the Ten Item Personality Inventory but none of the other covariates from the previous study were collected because they had already been found to be unrelated to the Knobe effect (e.g., OSPAN, SAT, CRT). We replicated the relation of extraversion with the Knobe effect in this national sample $r(294) = .20$, $p < .001$, observing a pattern of findings that was similar to the result from Cokely and Feltz (2009b) that was discussed above.

Theoretically, if extraversion is systematically and pervasively related to the Knobe effect, then extraversion should also predict the judgment asymmetry in structurally similar but different scenarios. If extraversion fails to predict in structurally similar but different scenarios, there may be something idiosyncratic about the Chairman Cases that is important for the relation to extraversion. In that case, the relation of extraversion to judgments in the Chairman Cases would not provide compelling evidence for general patterns of intentionally judgments. Does the predictive power of personality generalize across different Knobe-style cases?

To provide some additional evidence for the relation between extraversion and the Knobe effect, consider the second new study we conducted designed to test the extent to which extraversion remains robust across structurally identical but different Knobe-type assessments. The structurally similar scenarios described a Dean who harmed or helped qualified applicants as a side effect. One hundred and forty-seven participants were recruited from Amazon's Mechanical Turk panel of online workers. Thirty-two participants were excluded for not completing the survey or for requesting that their answers not be used. Fifty-two percent ($N = 60$) were male. Ages ranged from 18–76, $M = 30.91$, $SD = 11$.

Participants were randomly assigned to only one of two conditions. In the Harm condition, participants were given the Harmful chairman and a structurally similar Harmful Dean scenarios, counterbalanced for order. In the Help condition, participants received Helpful chairman and a structurally similar Helpful Dean scenarios, counterbalanced for order. The Harmful and Helpful Dean scenarios were as follows.

Harmful [Helpful] Dean

A professor at a university went to the Dean of the university and said, 'I want to start a new set of criteria for admissions into the university. It will help draw more attention to the university, but it will also harm *[help]* a lot of qualified and deserving applicants who do not have the funds for admissions.' The dean of the university answered, 'I do not care at all about harming *[helping]* qualified and deserving applicants from being admitted due to a lack of funds. I just want to make this university as recognized as possible. Let's start the new program.' They started the program and sure enough, qualified and deserving applicants without the funds were harmed *[helped]*.

Participants were then asked on a 7-point scale (1 = strongly disagree, 7 = strongly agree) to what extent they agreed with the following statement: "The dean intentionally harmed *[helped]* the applicants." After responding to the two scenarios, participants completed the Ten Item Personality Inventory and basic demographic information was gathered. In this study, we did not collect any additional covariates (e.g., CRT, OSPAN, SAT).

The Knobe effect was replicated for both the Chairman and Dean cases. Ratings of intentionality for Harmful Chairman ($N = 64$, $M = 5.59$, $SD = 1.71$) were significantly higher than intentionality ratings for Helpful Chairman ($N = 51$, $M = 2.04$, $SD = 1.59$) $F(1, 111) = 129.37$, $p < .001$, $\eta_p^2 = .54$. There was no main effect of order and order did not interact with judgments $Fs < 1$. Intentionality ratings for Harmful Dean ($N = 64$, $M = 5.61$, $SD = 1.56$) were significantly higher than intentionality ratings for Helpful Dean ($N = 51$, $M = 4.6$, $SD = 2.31$), $F(1, 111) = 6.31$, $p = .01$, $\eta_p^2 = .05$. However, this effect was qualified by an interaction of order. Judgments of Harmful Dean remained relatively stable when presented first ($M = 5.59$, $SD = 1.38$) compared to when it was presented second ($M = 5.63$, $SD = 1.72$), but Helpful Dean was markedly different when presented first ($M = 3.65$, $SD = 2.41$) compared to second ($M = 5.57$, $SD = 1.49$), $F(1, 111) = 10.39$, $p = .002$, $\eta_p^2 = .09$.

Correlations among the dependent variables for the helpful and harmful conditions are reported in Tables 3.1 and 3.2.

The relation of extraversion with Harmful Chairman was replicated and was the only personality trait reliably related to intentionality judgments. Neither sex nor age was related to intentionality judgments for Harmful Chairman. Largely the same pattern emerged for Harmful Dean. Extraversion trended toward significance and the size of the relation is

Table 3.1 Correlations for Helpful condition

	1	2	3	4	5	6	7	8	9
1. Dean	1	.19	.30*	.01	.1	.12	.1	.25^	-.11
2. Chairman		1	.11	.11	.16	.04	-.12	.07	-.03
3. Extraversion			1	.03	-.01	.09	.33*	.12	.06
4. Agreeableness				1	.11	.34*	.19	.1	.05
5. Conscientiousness					1	.24^	.06	.2	.16
6. Emotional stability						1	.17	.23^	0
7. Openness to experience							1	-.06	.24^
8. Age								1	.05
9. Sex									1

* $p < .05$, ^ $p < .1$

Table 3.2 Correlations for Harmful condition

	1	2	3	4	5	6	7	8	9
1. Dean	1	.69**	.22^	.12	.2	.27*	.18	.21^	.12
2. Chairman		1	.27*	-.03	.15	.04	.09	.15	.06
3. Extraversion			1	.22	.32*	.29*	.52**	.09	.05
4. Agreeableness				1	.03	.22^	.41**	.17	.29*
5. Conscientiousness					1	.42**	.08	.34**	.15
6. Emotional stability						1	.13	.13	-.15
7. Openness to experience							1	.03	.14
8. Age								1	-.01
9. Sex									1

** $p < .01$, * $p < .05$, ^ $p < .1$

consistent with previous studies. In the Helpful cases, extraversion was only related to Helpful Dean. This unexpected result may be partially due to the order effect present in the scenarios. Extraversion was not reliably related to Helpful Chairman.

To illustrate the relations between extraversion and intentionality judgments in the Harmful cases, rough extraversion quartiles were calculated. There were significant differences between extraverts (M = 6.69, SD = 0.48) and introverts (M = 5.14, SD = 2.07) in Harmful Chairman F (1, 23) = 5.1, p = .03, η_p^2 = .18. A similar difference was found between extraverts (M = 6.54, SD = 0.78) and introverts (M = 5.71, SD = 1.33) in Harmful Dean F (1, 23) = 6.06, p = .02, η_p^2 = .18 (see Fig. 3.2). Thus, these results suggest that the relations of extraversion to judgments about the side-effect effect are robust across different scenarios that employ harmful and helpful side effects.

We have emphasized the importance of replication in establishing relations among philosophically relevant intuitions and individual differences. To illustrate the importance of replication, some of our early work suggested that the order effect present with the Knobe effect was primarily attributable to responses of women (A. Feltz & Cokely, 2007). The somewhat surprising relation of the Knobe effect order effect with sex provides an illustrative example of some of the dangers of individual differences research (and empirical research in general), and serves as a reminder of how replications can help protect against common risks. In a third study, we attempted to replicate the finding that sex was associated with the order effect. This study also provided an additional attempt to replicate extraversions' relation to the Knobe effect (A. Feltz & E. T. Cokely, 2011).

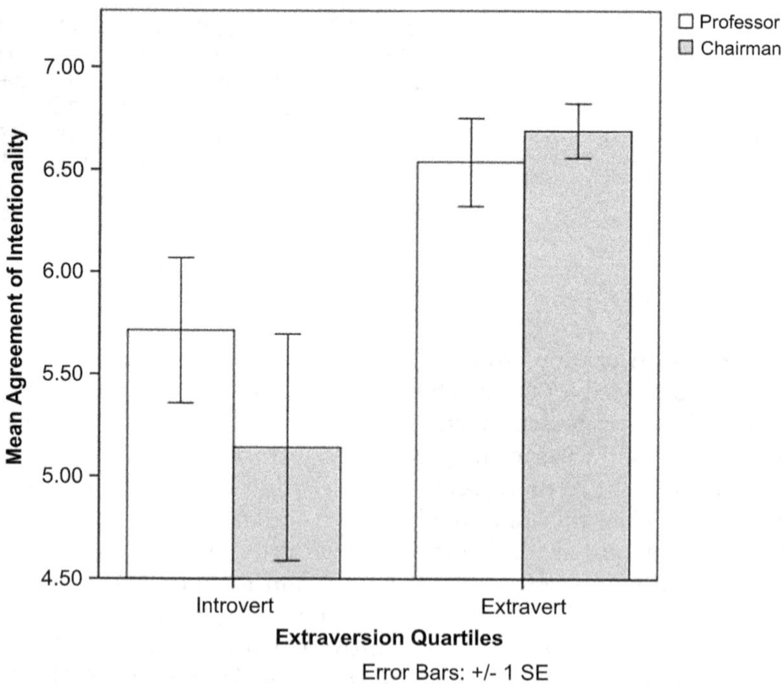

Fig. 3.2 Extraversion quartiles for Harmful Chairman and Dean

Again, participants responded to both the Harmful and Helpful Chairman Cases, counterbalanced (Cushman & Mele, 2008). Participants then rated their intentionality judgments about the side effects in each of those scenarios on a scale from 1–7 (disagree to agree). The results are reported in Table 3.3.

As the pattern of results above indicates, we observed the typical asymmetry in judgments.[4] When we conduced additional analyses, we replicated the order effect such that intentionally judgments for the Harmful Chairman were lower when Helpful Chairman was presented first compared to when Helpful Chairman was presented second.[5] But importantly, we failed to replicate the relation of the order effect with sex and sex was not otherwise related to the judgment asymmetry (Fs < 1).

The experiment also allowed for a replication of the relation of extraversion. In this analysis, we used a regression framework and included all of the Big Five personality traits, the CRT score, and sex as predictors and the judgment asymmetry as the outcome variable. The regression indicated a marginally significant relation of extraversion with the side-effect effect, presenting an effect size was consistent with previous studies.[6] Again, for illustrative purposes, we divided the participants in upper and lower extraversion quartiles. Using the quartiles as the independent variable indicated a near significant interaction with the judgment asymmetry.[7] In particular, those in the bottom quartile had lower intentionality judgments for the Harmful Chairman ($M = 4.57$, $SD = 2.3$) compared to those who were in the upper quartile ($M = 6.0$, $SD = 2.0$). Despite a

Table 3.3 Harm and Help responses by order

	The chairman intentionally helped the environment	*The chairman intentionally harmed the environment*
Help first	$M = 2.54$, $SD = 1.93$	$M = 3.90$, $SD = 2.27$
Help second	$M = 3.59$, $SD = 2.36$	$M = 6.01$, $SD = 1.63$

[4] $F(1, 89) = 46.55, p = .01, \eta_p^2 = .34$.
[5] $F(1, 87) = 4.03, p = .05, \eta_p^2 = .04$.
[6] $\beta = .21, t = 1.81, p = .07, R^2 = .12$.
[7] $F(1, 50) = 3.35, p = .07, \eta_p^2 = .06$.

reduction in power (due to reducing the number of participants in the analysis), extraversion was a significant predictor of harm judgments.[8] Extraversion did not reliably predict Help judgments ($F < 1$).

Although the experiment replicated the order effect, we did not replicate the relation of the order effect with sex. This may mean that sex is not normally an influential factor in the order effect or it could mean that sex is otherwise mediated by other more proximal factors (e.g., different distributions of personality between the sexes, etc.). However, the experiment replicated extraversion's relation to the Knobe effect. As such, this experiment serves as both converging evidence and corrective evidence, providing yet another cautionary tale for those conducting research focusing on individual differences or experimental manipulations.

The Knobe Effect, Extraversion, and Theoretical Accounts

The previous section suggests that extraversion may generally be tightly linked with the Knobe effect. But given the host of explanations offered in section "The Intentional Side-Effect Effect" of the Knobe effect, how does extraversion factor into these possible explanations? To illustrate one possible set of relations, extraversion could be related to the Knobe effect because extraversion (a) is mediated by some other concepts (e.g., extraverts may have a specific concept of intentional action where harmful, but not helpful, side effects are judged to be intentional or perhaps introverts have a different concept or set different thresholds for intentionality judgments compared to extraverts); (b) reflected an affective bias (e.g., extraverts may have an increased tendency to blame the Harmful Chairman more than the Helpful Chairman compared to introverts); or (c) both a and b. We conducted an experiment to help clarify the relations among a-c (Cokely & Feltz, 2009b). We assessed potential differences in concepts using a technique others have used where they presented some scenarios that did not generate or generated very low affective responses (Cushman & Mele, 2008; Nadelhoffer, 2006a). We attempted to test the affective biasing account by manipulation the order the scenarios were presented in. The reasoning goes like this. If non-affective scenarios were presented before potential affective scenarios, then one's core concept of intentional action might be activated and carry over from judgments about the non-affective scenarios to the affective scenarios, thereby reducing the overall

[8] ($M = 6.0$, $SD = 2.0$), $F(1, 50) = 5.32$, $p = .03$, $\eta_p^2 = .10$.

intentionally judgments in those affective cases. However, if the affective scenarios were presented before the non-affective scenarios and the affective biasing account is right, then we should see higher intentionality judgments in the affective cases compared to the alternate order. If that pattern of results is seen, then that would support, or at least be consistent with, an affective biasing account.

We conducted a study to test these different explanations of the side-effect effect. More importantly, we wanted to see if extraversion continued to predict the side-effect effect even after accounting for the multiple ways in which the effect could be produced. Specific individual differences in folk intuitions related to intentional action concepts were directly measured and manipulated (Cushman & Mele, 2008). In particular, the *belief-is-sufficient* concept was assessed with two scenarios. The first scenario we call Deer features a protagonist who accidentally kills another hunter as a side effect. Importantly, the protagonist did not believe he would kill the hunter and the protagonist also did not have a desire to kill the hunter.

> *Deer*: Imagine that there is a man out in the woods who is participating in a hunting competition. After spending hours waiting for a deer to cross his path, the hunter suddenly sees the largest deer he has ever seen. If he can only kill this deer, he will surely win the competition. So, the hunter gets the deer in his sights and pulls the trigger—thereby killing the deer. Unfortunately, the bullet exited the deer's body and struck a hunter who was hiding nearby. (Nadelhoffer, 2006a)

Participants then responded to the following prompt: "The man intentionally shot the hunter" on a 7-point scale that was used in the other studies (from disagree to agree; participants responded to the same Likert scale in all of the scenarios in this study). The purpose of this study was to highlight that sometimes there can be side effects that were not intended and not intentionally brought about.

The next scenario we called Eagle. The key features of this case were that the protagonist believes that the side effect will happen but does not want the side effect to happen.

> *Eagle*: Imagine that there is a man out in the woods who is participating in a hunting competition. After spending hours waiting for a deer to cross his path, the hunter suddenly sees the largest deer he has ever seen. If he can only kill this deer, he will surely win the competition. So, the hunter gets the deer in his sights—but at the last second, he notices that there is a beautiful

> eagle perched in a tree nearby. The hunter realizes that if he shoots the deer, the sound of the gunfire will definitely cause the eagle to fly away. But he does not care at all about the eagle—he just wants to win the competition. So, he shoots and kills the deer. And as expected, the sound of the gunfire causes the eagle to fly away. (Nadelhoffer, 2006a)

After reading the scenario, participants responded to the following prompt "The hunter intentionally scared away the eagle." Eagle allowed us to measure two different concepts of intention action. If participants responded that scaring away the eagle was intentional, then we classified those participants as having a belief-is-sufficient concept. If participants responded that scaring away the eagle was not intentional, then we classified those people as having a belief-is-insufficient concept.

Given our general theoretical perspective that most effects found in experimental philosophy (and behavioral science more broadly) are the predicable products of the interplay of person, process, and environmental factors, we predicted that many (if not all) of the factors identified above would contribute to the side-effect effect. In particular, we predicted that those who had the belief is-sufficient concept would have higher intentionality judgments than those who had the belief-is-insufficient concept because in both the Harmful and Helpful Chairman Cases, the chairman knew that the program would influence the environment. We would also expect that extraversion would predict the asymmetry. Finally, given the observed order effects in previous studies, we expected that there would be an order effect in this study that might support the affective biasing account (i.e., when one makes a judgment about a non-affective case, one may be more likely to use one's core concept and not be biased by other affective features of subsequent cases). The order effect should reduce the overall asymmetry when the non-affective cases are presented before the affective cases.

Since the predictions of this follow-up experiment were fairly complicated, they are summarized in the following points:

1. Priming the belief condition will result in an overall reduction of the intentional action side-effect asymmetry.
2. Those who are identified as having a belief-is-sufficient concept will judge both of the chairman's side effects as more intentional, as compared to those who have a belief-is-insufficient concept.

3. Extraversion will continue to account for unique judgment variance in the intentional action side-effect asymmetry after controlling for 1 and 2.

Executing the general strategy to link intuitions to personality traits, participants completed the Ten Item Personality Inventory and responded to Harmful Chairman, Helpful Chairman, Deer, Eagle. Here, we used two "blocks" to counterbalance the order. Participants received only one of the following two orders: [Deer, Eagle, Harmful Chairman, Helpful Chairman] or [Harmful Chairman, Helpful Chairman, Deer, Eagle]. Because we knew of the order effect that obtained between Harmful Chairman and Helpful Chairman, and to avoid a potential confound in the design, all participants received Harmful Chairman before Helpful Chairman.

Results concerning the Helpful and Harmful Chairman revealed the typical overall Knobe effect (Harmful Chairman M = 5.0, SD = 2.0; Helpful Chairman M = 3.2, SD = 2.0).[9] But things were more complicated after that basic finding. There were two higher-order interactions. The first involved those who were primed versus those who were not primed. Those who responded to the non-affective cases before the affective cases (Harmful Chairman M = 4.8, SD = 2.0; Helpful Chairman M = 3.4, SD = 2.0) had a significantly smaller judgment asymmetry between the Helpful and Harmful Chairman Cases than those who received the affective cases first (Harmful Chairman M = 5.5, SD = 1.9; Helpful Chairman M = 3.0, SD = 2.0, see Table 3.4).[10] The second interaction involved the two

Table 3.4 Intercorrelations for main variables

		1	*2*	*3*	*4*
1.	Extraversion				
2.	Belief-is-sufficient	.19*			
3.	Help	-.03	.29**		
4.	Harm	.18*	.54**	.28*	
5.	Side-effect asymmetry	.17*	.21*	-.60**	.60**

Notes: * $p < .05$; ** $p < .01$

[9] $F(1, 129) = 81.50, p = .001, \eta_p^2 = .38$.
[10] $F(1, 129) = 6.63, p = .01, \eta_p^2 = .05$.

concepts of intentional action measured in this study. Those who had the belief-is-sufficient concept had a larger judgment asymmetry (Harmful Chairman $M = 6.0$, $SD = 1.6$; Helpful Chairman $M = 3.6$, $SD = 2.2$) than those who had the belief-is-insufficient concept (Harmful Chairman $M = 4.2$, $SD = 2.0$; Helpful Chairman $M = 2.8$, $SD = 1.7$) (Fig. 3.3).[11] Additionally, those who had the belief-is-sufficient concept had higher ratings of intentionality for the side effects on both Harmful and Helpful Chairman.[12] We did not find evidence for a three-way interaction between concepts and priming conditions ($F < 1$).

Important for our purposes, we tested whether extraversion predicted the judgment asymmetry typical of the side-effect effect. To do so, we

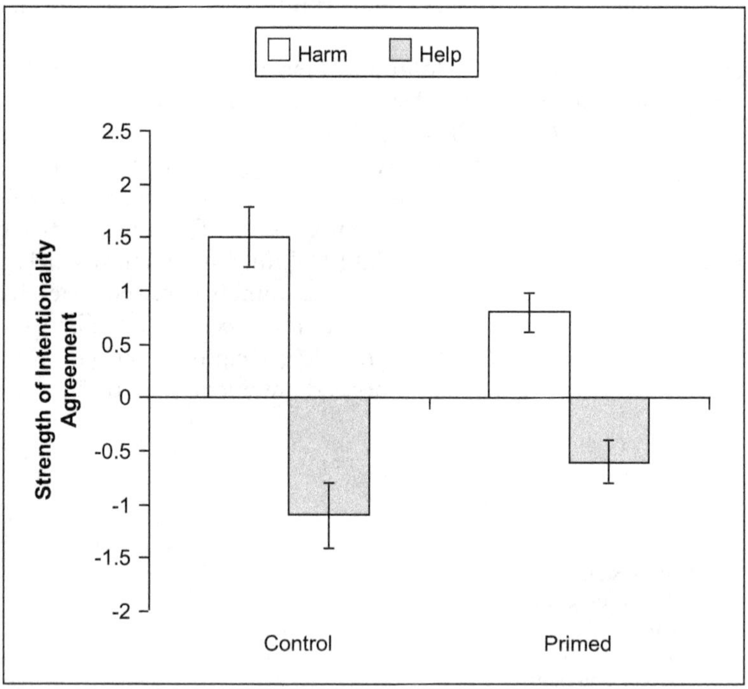

Fig. 3.3 Mean responses as a function of primed or not primed orders

[11] $F(1, 129) = 5.58$, $p = .02$, $\eta_p^2 = .04$.
[12] $F(1, 129) = 23.44$, $p = .001$, $d = .08$.

constructed a set of hierarchical linear regressions. Hierarchical regressions proceed by including predictor variables in a specific order and then determining if subsequent predictor variables improve model fit. We included each of the relevant independent variables as predictors in three different models. The final model had three independent variables including (1) priming, (2) extraversion, and (3) concepts as predictors of harm judgments (Table 3.5). This full model accounted for a large amount of variance in judgment.[13] After controlling for priming, extraversion remained a reliable predictor of the harm judgments. Thus, part of the predictive power of extraversion appears to result from extraversion's positive association with the belief-is-sufficient concept. However, extraversion's effect was also mediated by the large effect of concepts (Table 2.2).

Subsequent analyses were performed to evaluate hypothesis 3. Hierarchical regression models examined extraversion within each of the two concept groups. Regression analysis indicated that for the belief-is-sufficient group, extraversion was unrelated to judgment asymmetry ($F < 1$). Regression analysis next assessed the belief-is-insufficient group using (1) priming and (2) extraversion as predictors. This model was a reliable predictor of the judgment asymmetry.[14] Consistent with Hypothesis 3,

Table 3.5 Hierarchical linear regression analysis explaining intentional action judgments in the Harm condition

Steps and variables	*Beta*	*R*	R^2	ΔR^2	*F*
Model 1.					
Order-effect	.13	.13	.02	.02	2.43
Model 2.					
Order-effect	.12				
Extraversion	.17*	.22	.05	.03	3.18*
Model 3.					
Order-effect	.18*				
Extraversion	.06				
Belief-is-sufficient	-54**	.58	.33	.28	21.37**

Notes: * $p < .05$; ** $p < .01$

[13] $F(3, 130) = 21.37, p = .001, R^2 = .33$.
[14] $F(1, 67) = 3.29, p = .04, R^2 = .09$.

after controlling for the effect of priming, extraversion continued to account for unique variance for those individuals who had the belief-is-insufficient concept.[15] Additionally, the unique effect of extraversion was found to primarily reflect a relationship with harm judgments (Figs. 3.4 and 3.5).[16]

The results of this experiment provide support for Hypotheses 1–3. As predicted by Hypothesis 1, the judgment asymmetry typical of the Knobe effect was smaller following priming. This provides causal evidence of the presence of judgment bias in both Harmful and Helpful Chairman (see also Nadelhoffer (2004)). However, the priming effect was relatively small

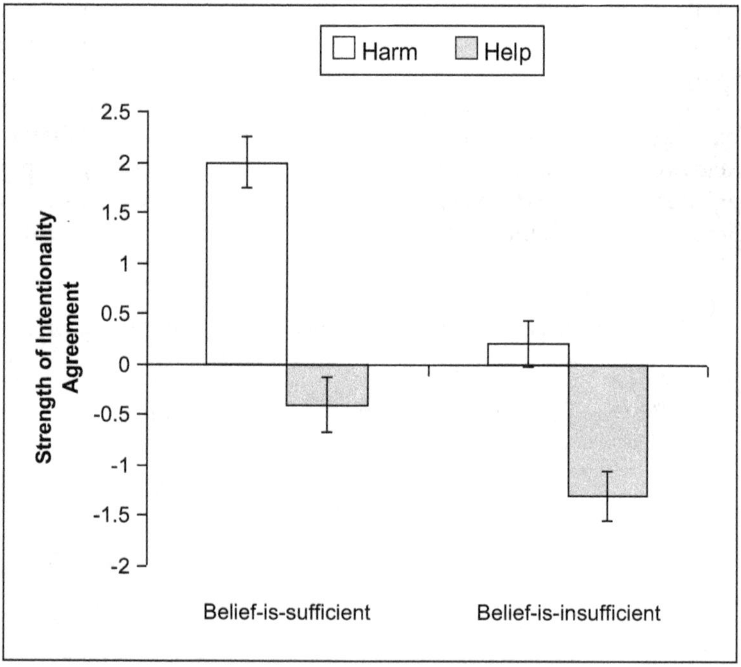

Fig. 3.4 Large, qualitative differences in the judgment asymmetry are predicted by individual differences in specific concepts (belief-is-sufficient, belief-is-insufficient). Positive numbers indicate agreement, negative numbers indicate disagreement, and error bars represent the standard error of the mean

[15] $F(1, 67) = 4.74$, $p = .03$, $R^2_{change} = .06$.
[16] $r(68) = .24$, $p = .05$, (Helpful Chairman n.s.).

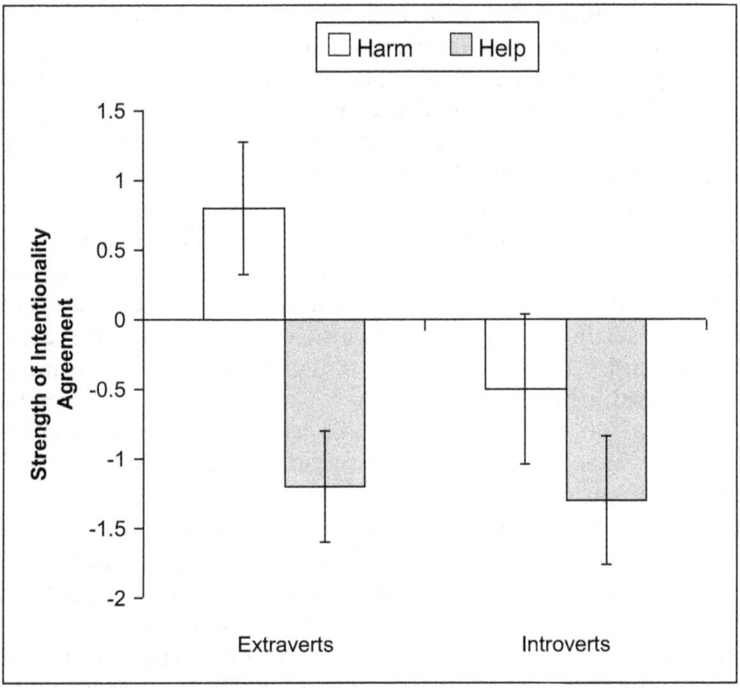

Fig. 3.5 The side-effect judgment asymmetry (Help versus Harm) for belief-is-insufficient, which is predicted by extraversion (Introvert versus Extravert). Positive numbers indicate agreement, negative numbers disagreement, and error bars represent the standard error of the mean

compared to the large, qualitative judgment differences associated with different concepts. Consistent with Hypothesis 2, we observed that people with a belief-is-sufficient concept tended to judge actions as more intentional whereas people with a belief-is-insufficient tended to judge those actions as unintentional. Moreover, although it was not predicted, extraversion was a significant predictor of differences in concepts. Extraverts were more likely to belong to the belief-is-sufficient group whereas introverts were more likely to belong to the belief-is-insufficient group. Finally, extraversion predicted unique variance even when controlling for different concepts.

These experiments suggest that a variety of factors play theoretically important roles in intentional action judgments. The intentional action side-effect asymmetry does not appear to result from a single mechanism but instead reflects robust influences of both individual differences and judgment processes (e.g., affective biases). Previous experiments provided evidence of individual differences because extraversion was strongly related to the side-effect asymmetry even after controlling for other potentially influential individual differences (e.g., cognitive abilities, sex). The current experiment provided converging evidence of a judgment bias demonstrating that priming intentional action concepts causally reduced the judgment asymmetry. The relation between extraversion and the judgment asymmetry was replicated in the current study; however, extraversion's effect was explained in part by its association with specific concepts. Extraverts tended toward a belief-is-sufficient concept whereas introverts tended toward a belief-is-insufficient concept. These specific individual differences in concepts were in turn associated with large, qualitative and theoretically important differences in judgment. Those who had the belief-is-sufficient concept tended to judge that all side effects were more intentional whereas those who had a belief-is-insufficient concept tended to judge that all side effects were less intentional. Finally, when individual differences in concepts were taken into account, extraversion continued to predict unique variance for harm judgments for some people (i.e., belief-is-insufficient group). In summary, the Knobe effect appears to be generally a function of the interplay of the main factors: (1) affective judgment biases, (2) specific concepts, and (3) personality.

Framing Intentional Action Intuitions

Extraversion predicts philosophical bias for some paradigmatic and theoretically important intentional action intuitions—namely, the Knobe effect. One explanation of the relation between extraversion and the Knobe effect is that extraverts may have different sensitivities or motivations than introverts. Extraverts may be more sensitive than introverts to the socially undesirable elements in the Harmful Chairman case and may be more motivated to express this. If extraverts are differentially sensitive than introverts to some of the elements in the Chairman Cases, then it stands to reason that one should be able to change the nature of the relation between extraversion and intuitions reflecting the Knobe effect. This can be done by either changing elements of the scenarios or by changing the environment in which people are asked to make decisions.

To show that intentional action intuitions can be predictably manipulated by changing key elements of the scenarios, we gave participants revised versions of Harmful and Helpful Chairman. These revised versions took inspiration from classic framing cases (Tversky & Kahneman, 1981). These classic cases have been used to demonstrate that the way that options are described can alter the choices that many individuals end up making, even if the options are apparently logically identical. In these cases, a new disease puts 600 people at risk of dying. In one condition (the negative frame), people can choose between adopting Program A where 400 people will die for sure or Program B where there is a 1/3 chance of nobody dying and a 2/3 chance of everybody dying. When described in terms of dying, around 20% of people chose program A. In a separate condition (the positive frame), people make a similar choice except the wording is changed. Instead of 400 people dying in Program A, they are told that Program C will save 200 people. Program D is described as having 1/3 chance of saving everybody and a 2/3 chance of saving nobody. Remarkably, in the saving condition, 70% of people chose Program C. Critically the expected utility of all the programs is the same. One could expect that on average, 400 people would die and 200 would live in each of the choice tasks. In terms of expected utility, the two choice tasks are identical. However, the "frame" of the choice influenced judgments of some people.

In this light, the Harmful Chairman was revised so that the side effect was framed positively (saving rain forest) even though the side effect harms the environment. Helpful Chairman was likewise altered so that the helpful side effect is framed negatively (destroying rain forest) but the environment is helped. One hundred and sixteen undergraduates at a large state university were recruited to participate in the study. One group of participants ($N = 58$) received the following scenario.

Chairman Help Negative

The vice-president of a company went to the chairman of the board and said, "We are thinking of starting one of 2 new programs. Either one will help us increase profits for this year's balance sheet, and in ten years they will help the environment by affecting up to 6000 acres of a rain forest. If we choose program A, 4000 acres of the forest will be destroyed with certainty. If we choose program B, there is a 67% chance that 6000 acres will be destroyed." The chairman answered, "I don't care at all about the trees of the rain forest. I just want to make as much profit for this year's balance sheet as I can. Let's start program A." They started the new program. Sure enough, in ten years the rain forest was helped.

A separate group of participants ($N = 58$) received the following scenario.

Chairman Harm Positive
The vice-president of a company went to the chairman of the board and said, "We are thinking of starting one of 2 new programs. Either one will help us increase profits for this year's balance sheet, but in ten years they will harm the environment by affecting up to 6000 acres of a rain forest. If we choose program A, 2000 acres will be saved with certainty. If we choose program B, there is a 33% chance that 6000 acres will be saved." The chairman answered, "I don't care at all about the size of the rain forest. I just want to make as much profit for this year's balance sheet as I can. Let's start program A." They started the new program. Sure enough, in ten years and the rain forest was harmed.

Participants were then asked to rate their agreement with the appropriate version of the following statement (on a 7-point scale, 1 = strongly disagree, 4 = neutral, 7 = strongly agree): "The chairman intentionally harmed/helped the rain forest." The additional information about risk describes logically identical options—on average there would be 4000 fewer acres of rain forest. However, the description of "saving" and "destroying" theoretically should alter some people's judgments about the intentionality of the side effect. Specifically, if extraverts are more sensitive to socially unacceptable consequences or are motivated to act on those consequences, destroying the rain forest should impact their intentionality judgments more than introverts even if the side effect ultimately helps. Similarly, in the Harm case, when the action is described as saving the rain forest, extraverts should be more likely than introverts to judge the action as less intentional even if the side effect is harmful. However, introverts should be relatively less affected by this additional information and give responses similar to pervious experiments (i.e., display a muted Knobe effect). After reading one of the scenarios, participants filled out the Ten Item Personality Inventory.

An ANOVA indicated a statistically significant, but markedly reduced, Knobe effect between the Chairman Harm Positive ($M = 3.88$, $SD = 1.73$) and Chairman Help Negative ($M = 2.29$, $SD = 1.57$).[17] This muted Knobe effect appears to be the result of the framing. We observed the predicted reversal in extraversion's relation to judgments about the harmful and helpful side effects. When the harm was framed positively (i.e., saving

[17] $F(1, 114)$, $p < .001$, $\eta_p^2 = .19$.

parts of the rain forest), extraversion was negatively related to intentionality judgments, r (58) = −.27, p = .04. Likewise, when the help brought about was framed negatively (destroying parts of the rain forest), extraverts were more likely to judge the help to have been brought about intentionally r (58) = .33, p = .01. To illustrate further, rough quartiles were constructed for extraverts and introverts. Extraversion had a statistically significant interaction effect with judgments about Chairman Harm Positive (Extraverts: N = 13, M = 3.69, SD = 1.84, Introverts: N = 13, M = 4.69, SD = 1.65) and Chairman Help Negative (Extraverts: N = 14, M = 3.21, SD = 2.01, Introverts: N = 15, M = 1.73, SD = 1.39)[18] (see Fig. 3.6).

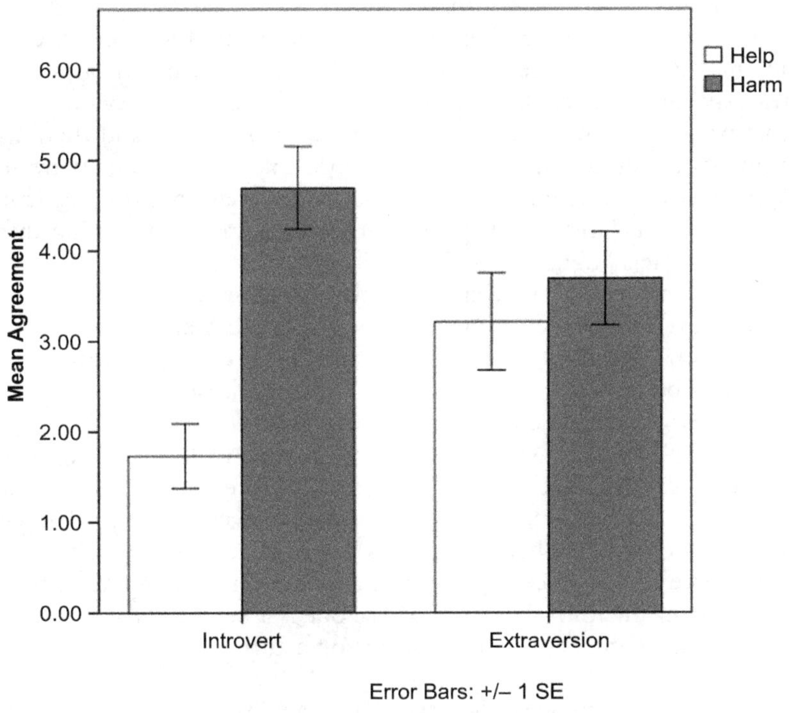

Fig. 3.6 Intentional action framing by extraversion

[18] $F(1, 51) = 7.04$, $p = .01$, $\eta_p^2 = .12$.

As this study suggests, the relation of extraversion to judgments about side effects can be predictably manipulated. In this case, there is something about losses that extraverts seem to be especially sensitive to. When the emphasis on losses is minimized, extraverts have a different set of intuitions. As such, intuitions about the Knobe effect are partially a result of the decision task along with personality. In the next section, we detail an experiment that indicates that intuitions about the Knobe effect are also partially a result of the affordances in the decision making environment—i.e., the choice architecture.

Perspective in Intentional Action Attributions

To take stock, some people judge the intentionality of some bad side effects differently from some good side effects. Extraversion predicts some of these differences. Moreover, intuitions about the Knobe effect can be manipulated, and the manipulation can be predicted based on some of extraverts' sensitivities. One common feature of the studies reviewed so far is that participants were asked to imagine some scenarios and then make judgments about them. However, it seems likely a different pattern of judgments would be found if participants were actually performing actions with side effects. Indeed, there is a large literature documenting differences in people's judgments and behaviors as a function of whether one is performing an action (i.e., an actor) versus observing the behavior of another (i.e., an observer) (Jones & Nisbett, 1972; Malle, 2006; Malle & Knobe, 1997; Malle, Knobe, & Nelson, 2007). Just imagine driving a car when somebody cuts you off. What do you think of that person? Now imagine that you do exactly the same thing and cut someone off. Are your thoughts about your own versus the other person's behaviors different? The literature suggests that they would be. For example, you would likely have a much harsher judgment of the person who cut you off (that jerk!) compared to your own behavior (I was just in a hurry!).

The same basic principles have been applied in experimental philosophy. Consider the following cases where one is asked to make a judgment as a third-party observer or as a first-person actor:

> A trolley is hurtling down the tracks. There are five workers on the track ahead of the trolley, and they will definitely be killed if the trolley continues going straight ahead since they won't have enough time to get out of harm's way. There is a spur of track leading off to the side where another person is

working. The brakes of the trolley have failed and there is a switch which can be thrown to cause the trolley to go to the side track. Imagine that you are an innocent bystander who happens to be standing next to the switch. You realize that if you do nothing, five people will definitely die. On the other hand, you realize that if you throw the switch, you will definitely save the five workers. However, you are also aware that in doing so the worker on the side track will definitely be killed as the result of your actions. (Nadelhoffer and Feltz (2008) see also Petrinovich and ONeill (1996))

As the cases illustrate, the only major difference between the scenarios was whether judgments were made about John or about "you." The key dependent variables in this scenario were whether it was permissible to throw the switch and how much control one had over the situation. Consistent with the actor-observer literature, people tended to think it more permissible for John to throw the switch (90% said permissible) and that John had more control ($M = 3.72$) than when they were given the "you" version (65% and $M = 2.88$). These results indicate that one's viewpoint can influence judgments about moral permissibility and control (Nadelhoffer & Feltz, 2008). This finding is consistent with a large literature that differences in perspective can alter some judgments including reasons for performance on tests (Nisbett & Wilson, 1977), reasons for choices of academic major or girlfriends, and reasons for volunteering (Nisbett, Legant, & Marecek, 1973) (see for reviews, Baron and Branscombe (2012); Malle (2006)).

Given the pervasive impact of actor-observer differences, these differences would also likely exist for some intentional action intuitions. One way to illustrate the actor-observer difference with respect to the Knobe effect is to take a similar strategy used in the Trolley case above—simply ask one group to imagine they are the chairman and compare response to another group that responds to the normal Chairman scenarios. However, when participants were asked to imagine being the chairman, the same patterns of results typical of the Knobe effect were found and not reliably different from the original cases where one was an observer. A stronger manipulation than asking people to imagine being an actor or an observer is likely needed. Feltz, Harris, and Perez (2012a) thought that one way to generate these effects was not just to ask people to imagine being the chairman but to put them in a situation where they actually were (like) the chairman. That is, they thought that participants needed to take some real action and make a decision that had some good or bad outcome to be like the chairman.

To allow for participants to actually become actors, Feltz, Harris, et al. (2012a) had participants play a "game." The game was the kind of game that is typical for an area of research called Behavioral Economics where people made decisions about allocating resources against other people in the same game. In this game, the decisions that the players make impact other players. How well a player does in the game depends both on the individual decisions of the player along with the decisions of all the other players in the game. In Behavioral Economics, "how well" a player does is often determined by how many "experimental currency unites" (or ECUs) a player accrues in the game. Then, at the end of the game, the players trade in their ECUs for *real* money. So, there is no lip service in this game since the rewards (and punishments) were real and not hypothetical.

The game that the players played was as follows. Participants came to a computer lab in 4 different groups of between 4 and 12 people (total number of participants was 45). Participants were instructed about the rules of the game before starting to play. They were told that they would earn $10 for showing up to the study (a standard at the time in Behavioral Economics studies) and they would have the chance to earn more money depending on how they and others played (as a matter of fact, final payoffs were between $16 and $20). After getting these instructions, participants were assigned to one of four conditions. Participants were assigned to a Help and a Harm condition. In the Harm condition, participants were instructed that they had ten tokens that they had to decide what to do with. They only had two options: they could keep all their tokens, or they could contribute any percentage of those tokens (i.e., 1–10) to a "group account." For each token they kept, they would get 10 ECUs toward their final reward at the end of the experiment. However, here's the rub. If the player invests a token in the "group account," that player would earn 12 ECUs but all the other players in the game would receive a 3 ECU penalty. The other condition was the Help condition. In the Help condition, the instructions were exactly the same as they were in the Harm condition except that contributing to the group account would generate a 3 ECU bonus to others in the game. The final condition was the Observer condition. In this condition, participants did not have to take any action but read about the action another person took that impacted them (i.e., they observed the game playing action of another person). In this case, either the other person's action generated a bonus or penalty in line with the

Harm and Help conditions. In this case, Feltz et al. (2012a) simply set things up so that somebody was described as contributing ten tokens to the group account thereby generated a 30 ECU bonus or penalty. Hence, given the setup of the game, actor-observer differences might be observed not for hypothetical but for real cases.

The four conditions that participants were exposed to were determined as follows. People were both actors and observes but only in one of the Help or Harm conditions. The four conditions were generated by counterbalancing the orders of being an actor or an observer. After each instance of observing or acting, participants responded to the key 3 statements in this study (the actual contribution to the accounts was not the main dependent variable in the study):

1. "*You/the other participant* intended to generate the *penalty/bonus.*
2. *You/the other participant* intentionally generated the *penalty/bonus.*
3. *You/the other participant are/is blameworthy/praiseworthy* for generating the *penalty/bonus.*"

Participants responded on a 7-point Likert scale (1 = disagree, 7 = agree).

Feltz, Harris, et al. (2012a) were only interested in looking at responses to 1–3 from people who performed some action. It was simply an unavoidable element of the experiment that some people might decide *not* to contribute to the group account in which case they did not perform an action to contribute. So, they only looked at participants who contributed to the group account. Overall, 18 people made a decision to contribute to the group account in the Harm condition and 20 people made a decision to contribute to the group account in the Help condition. The responses to prompts 1–3 indicated that there was indeed an actor-observer asymmetry in judgments (see Table 3.6).

So far, these results suggest that there can be actor-observer differences with intentional action intuitions. But these overall intuitions do not

Table 3.6 Means and standard deviations for actors' and observers' intention and intentionality judgments

Actor intend	$M = 3.26$ $SD = 2.13$
Actor intentional	$M = 3.5$ $SD = 2.20$
Observer intend	$M = 4.16$, $SD = 2.1$
Observer intentional	$M = 4.26$, $SD = 2.17$

necessarily inform actor-observer differences associated with the Knobe effect. The harm/help to the other participants may not be *side effects* of the intended action to make greater profits. To find those who viewed the harmful or helpful consequence as a side effect, those who reported that the harm or help was not intended were identified (those answering 1–4 on the intention question above). After these exclusions, 16 participants did not intend the harmful outcome of investing and 10 participants did not intend the helpful outcome. As predicted, Feltz, Harris, et al. (2012a) found an actor-observer difference with the side effects. In the Actor condition, the harmful outcomes were thought to be less intentional (M = 2.31) than the helpful outcomes (M = 3.1). However, there were no reliable differences in the Observer condition

A question that was centrally important for the purposes of this chapter is whether extraversion was related to the asymmetry in judgments. A similar strategy was taken in this study and those reviewed above. Participants also completed the Ten Item Personality Inventory. Here, extraversion was strongly correlated with the harm judgments for those in the Actor condition, $r = -.55$, $p = .03$, and was not significantly correlated with actors' intentional help judgments $r = .40$, $p = .25$. To help further understand those relations, Feltz, Harris, et al. (2012a) created two groups based on extraversion scores. Those two groups were identified by those who were in the top half of extraversion versus bottom half of extraversion (i.e., a median split). Using the median split as an independent variable along with the Harm/Help condition resulted in the expected interaction (see Fig. 3.7).

Meta-Analysis

Given the number of studies reported in this chapter, it is useful to have an overall summary of the relation between extraversion and the Knobe effect. So, we conducted a meta-analysis to combine the results reported in this chapter.

Some of the effects reported in this chapter were predicted to have the opposite sign. For example, it was expected in the framing cases (i.e., framing the Harmful chairman positively), the effect of extraversion would be opposite of the normal direction of the relation of extraversion with the Harmful chairman. Because these differences were predicted, we accounted for this difference by changing the direction of the signs so that they

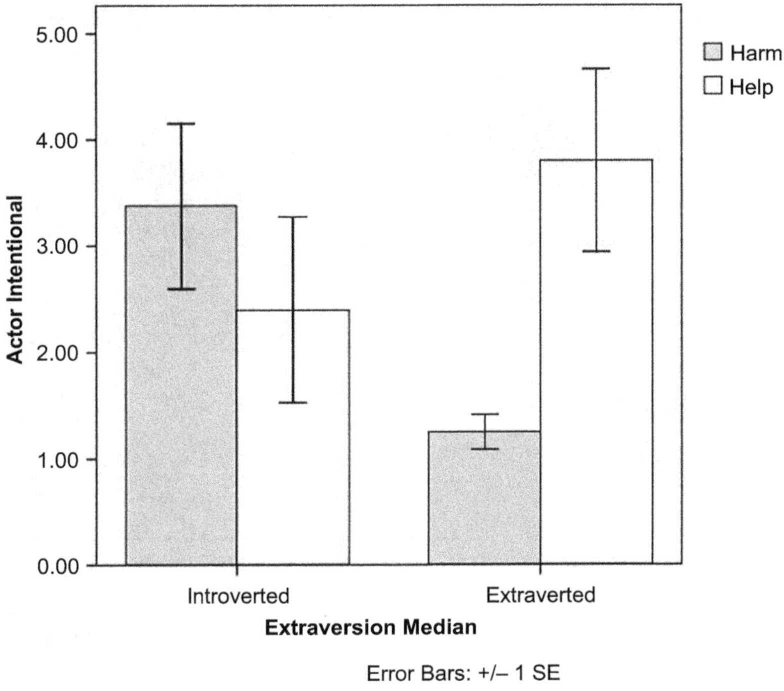

Fig. 3.7 Extraversion median intentionally means. Error bars represent 1 standard error

would be consistent with the predictions of extraversion's relation to the Knobe effect (i.e., we made all the predicted relations positive).

Given these stipulations, the overall mean effect size between extraversion and judgments typical of the Knobe effect was estimated to be .24, (95% *CI* 0.17–0.31), $p < .001$ (see the forest plot in Fig. 3.8). No heterogeneity was observed in the effect sizes $Q(7) = 3.72$, $p = .82$, suggesting that the relation remained stable regardless of the experiment. A test of the funnel plot did not reveal evidence of publication bias, $z = 1.64$ $p = .1$, but this result should be interpreted with caution since we did not do a complete, full search for any unpublished studies. Hence, there is reason to be quite confident that the relation between extraversion and the Knobe effect is robust across several different testing environments and materials.

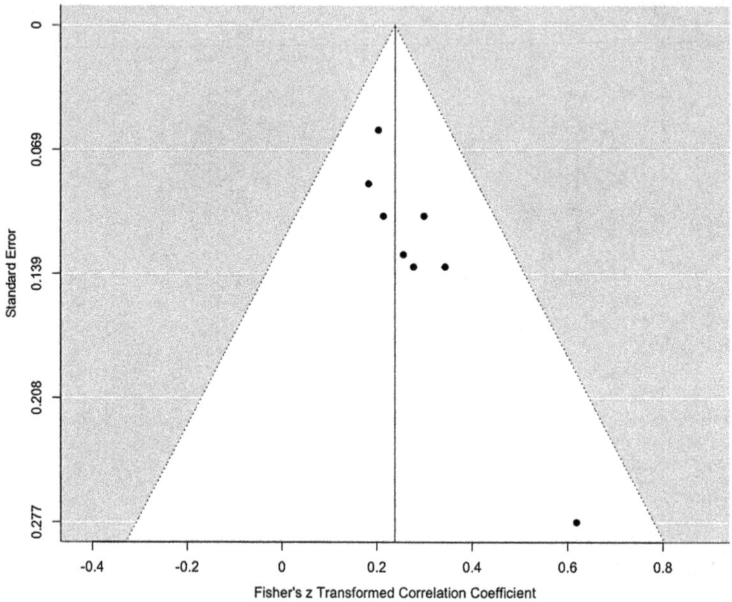

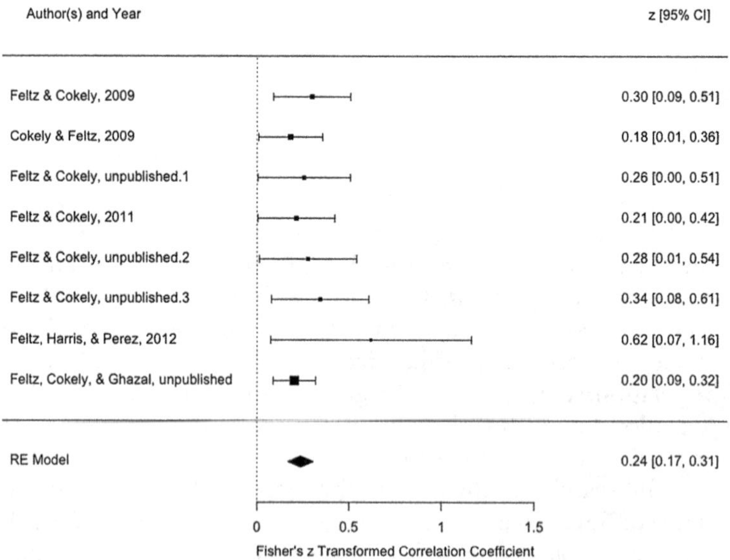

Fig. 3.8 Forest plot of the relation between extraversion and the Knobe effect

Conclusion

In this chapter, we have surveyed a number of different studies suggesting that extraversion is systematically related to an important class of intentional action intuitions about side effects. The variety of experiments suggests that there is not something idiosyncratic about the classic Chairman Cases that accounts for the relation between extraversion and intuitions about side effects. Changing the nature of the scenarios predictably changed the relation of extraversion with intentionality judgments. Similarly, changing the decision making environment predictably altered the relation of extraversion to intentional action intuitions. The meta-analysis suggested that the relation remained consistent across these different experiments. In summary, experimental data and statistical modeling indicate that many intuitions relevant to the Knobe effect are systematically related to some general, heritable personality traits.

The persistent relation of individual differences to the Knobe effect challenges many explanations of the Knobe effect. Almost all the models concerning the Knobe effect are based on mean responses and do not take into account individual differences. Additionally, most of the theories at least tacitly hold that there is one factor that can account for the Knobe effect. Even at the broadest level of explanation (i.e., core concept v. non-concept explanations), the notion that a single explanation for the side-effect effect can account for the varied responses is just wrong.

The data we have reviewed present a substantially different picture of what would be required for an adequate account of the Knobe effect. There is not necessarily any intentional action judgment asymmetry as there may not typically be any complete judgment reversal. Rather than any single general "bias" or judgment process that causes participants to switch from intentional (harm) to unintentional (help) judgments, the change in judgment appears to be more modest, changing for example from neutral judgments to intentional or from unintentional to neutral. The only identifiable and somewhat complete judgment reversal involved the group of extraverts who behaved as if they held a belief-is-insufficient concept. Additionally, we found that by changing the task (framing) or the judgment environment (making people actual actors) we could predictably change the overall pattern of intuitions. In these ways, the observed judgment asymmetry, which seems to be the product of the interplay of several distinct mechanisms, may be more accurately characterized as a collection of intentional action biases or multiple judgment asymmetries.

If all of this is right, then our results indicate that a parsimonious account of folk intentional action intuitions is not likely forthcoming. Intentional action intuitions appear to be related to a number of independent factors including one's personality, biases, the task, and the task environment. Because there is no single judgment asymmetry, the intentional action intuitions generated typical of the Knobe effect seem to indicate that there is no *the* folk concept of intentional action. Once again, the evidence suggests there is no single, monolithic folk concept of intentional action. Rather, there appear to be several. We expect that there will be "*groups* of folk" who express different, stable, predictable, and philosophically interesting intuitions (Cushman & Mele, 2008). Hence, proposing a theory or conceptual analyses based on the presupposition that there is only one folk concept, set of intuitions, or mechanism will be incomplete or inaccurate.

It is still *possible* to propose a single theory or conceptual analysis in light of the different factors that are related to intentional action intuitions. But possibility is cheap. The mounting evidence suggests that there is stable diversity in people's intentional action intuitions. If there is predictable and stable variation of intentional action intuitions, then a theory or conceptual analysis concerning intentional action that takes those intuitions as evidence must account for that variation in some way. For example, theorists may treat the variation as confirming evidence (e.g., relativism, conceptual diversity), or the theorist may explain why at least some of the intuitions are wrong (e.g., an error theory). In any event, an account of those intuitions is required for most theories or conceptual analyses about intentional action. Simply holding that it is possible to give such an account is insufficient to deflect the worries presented by stable individual differences in intuitions. Therefore, the philosophical use of intentional action intuitions *requires* a comprehensive understanding of the extent of variation in those intuitions.

In conclusion, the evidence presented in this chapter provides support that an important class of intentional action intuitions is related to a global personality trait. In the previous chapter, we have seen that some free will intuitions are related to one's personality. Consequently, there is gathering evidence that personality is meaningfully related to many philosophically relevant intuitions and that the relation is not limited to one kind of intuition (e.g., intuitions about free will). In the next chapter, we review another set of philosophically relevant intuitions related to personality—ethical intuitions.

CHAPTER 4

Ethics

"Imagine that John and Fred are members of different cultures, and they are in an argument. John says, 'It's okay to hit people just because you feel like it,' and Fred says, 'No, it is not okay to hit people just because you feel like it.' John then says, 'Look you are wrong. Everyone I know agrees that it's okay to do that.' Fred responds, 'Oh no, you are the one who is mistaken. Everyone I know agrees that it's not okay to do that.' (Nichols, 2004a, pp. 9–10)
Suppose somebody asks you who is right in the debate? Is Fred right that it is not okay to hit people just because you feel like it? Is John right that it is okay to hit people just because you feel like it? Or is there no fact of the matter about claims like hitting others just because you feel like it?"

The thought experiment presented above is designed to assess intuitions that are central to people's beliefs about moral objectivism. Moral objectivism, as we will use the term, is the view there are some moral statements that are true or false independent of what anyone thinks about the contents of those statements (Mackie, 1977). In simple terms, while some people think morality is relative to what people think about those issues, most moral objectivists think that some things are just clear-cut right or wrong regardless of one's situation, culture, or values. Debates about moral objectivism have been central parts of contemporary ethics for thousands of years. As will come as no surprise to the reader by now, there is persistent debate about whether moral objectivism is true. These

A. Feltz, E. T. Cokely, *Diversity and Disagreement*,
https://doi.org/10.1007/978-3-031-61935-9_4

disagreements, at least in part, are dependent on the intuitions that we express and take as evidence about cases illustrating some aspect of moral objectivism. Could personality be related to these fundamental intuitions and corresponding debates about moral objectivism?

Taking an even broader perspective, the focus of this chapter will be on ethics, personality, and morally relevant behavior. This will be the last chapter where we provide a detailed review of empirical findings on the relations between personality and philosophical intuitions before moving onto some related theoretical, philosophical, and practical implications of these empirical findings. Because ethics is such a broad field, we will limit the scope of our review to a handful of fundamental, theoretically pivotal issues with broad ramifications.

Specifically, we will focus on research demonstrating that personality predicts intuitions relevant to meta-ethics, first-order ethics, and applied ethics, as well as predicting actual morally relevant behaviors and outcomes. As a result, the evidence presented in this chapter spans broader areas of ethics rather than narrowly focusing on just a few philosophical issues. This breadth may result in an impression that the relations between ethical intuitions and personality are more fragmented and not as thoroughly investigated, which could further give the impression that the evidence is in some ways less convincing than findings presented in the past two chapters. We agree this is a noteworthy difference for many reasons (e.g., replication in science can be valuable and necessary). Nevertheless, we hope that the relative lack of intensive focus in this chapter is compensated for by evidence on the considerable breadth of the associations between personality and intuitions about ethics. In any event, the data in this chapter are consistent with our central position that personality traits are often robustly and systematically linked with a number of philosophically relevant judgments. We are also happy to admit that there is still plenty of work left to be done on these and many other issues, and we hope that others will continue to explore the relations discussed here and throughout this book.

Personality Predicts Meta-Ethics

Meta-ethics is one prominent area in ethics. Meta-ethics largely deals with questions *about* ethics rather than attempting to determine correct substantive theories about morally right or wrong actions. For example, if you have moral objectivist tendencies, you probably thought that either John or Fred was right in the scenario presented at the beginning of this chapter. According to the moral objectivist, if needless suffering is bad, it is bad regardless of what anyone thinks about that suffering. Needless suffering is bad even if nobody can understand or think about the suffering. Many contemporary philosophers think that some form of moral objectivism is true (Shafer-Landau, 2003; M. Smith, 1995). Not only do some philosophers think that moral objectivism is true, they also think that moral objectivism is supported by and is deeply entrenched in everyday thought about morality. A belief in moral objectivism has been argued to be essential to moral cognition, the regulation of interpersonal relationships, and the prevention of moral nihilism (e.g., a belief that ultimately nothing is morally right or wrong) (Lycan, 1986; Mackie, 1977). If people were to give up their belief in moral objectivism, it is argued that life would lose the deeper meaning, satisfaction, and purpose it once had (C. Wright, 1992). Like free will, some have argued that if we find that moral discourse is deeply flawed in its commitment to moral objectivism, we should leave people to their mistaken beliefs. To correct those erroneous beliefs would have unwanted and dire consequences (Joyce, 2001).

All of this assumes that people have a belief in moral objectivism. However, empirical data suggest a substantial number of people have non-objectivist intuitions. Nichols (2004a) found that many people expressed non-objectivist intuitions about a canonical moral violation (i.e., harming another person just for fun). Theoretically, out of all kinds of moral violations, harming another person for no good reason should have a strong claim to objective truth. According to moral objectivism, if hitting another person just for fun is morally wrong (or right), it is simply wrong (or right). It is not possible for hitting another person for fun to be morally right for some people and morally wrong for other people, everything else being equal. As such, moral objectivist intuitions can be operationalized by assessing whether one thinks that it is possible for two people in a moral disagreement to both be correct.

Moral objectivism is another philosophically complicated notion (like determinism, side effect, and intentional action). As illustrated in previous chapters, theorists often create scenarios to illustrate central features of philosophically complicated concepts to help non-experts understand those features. In this case, theorists have created scenarios to capture key elements of moral objectivism to test folk intuitions. Take another look at the scenario you read at the beginning of this chapter. We'll call this scenario *Moral*. Moral is one scenario that theorists have created to test objectivist intuitions. If one responds that either John or Fred but not both are correct, then one expresses objectivist-friendly intuitions (responding #1 or #2 below). However, if one thinks that John and Fred are both correct, then one expresses non-objectivist friendly intuitions (responding #3 below).

Nichols (2004a) gave participants Moral and asked them to indicate which of the following best characterizes the nature of the disagreement:

1. It is okay to hit people just because you feel like it, so John is right and Fred is wrong.
2. It is not okay to hit people just because you feel like it, so Fred is right and John is wrong.
3. There is no fact of the matter about unqualified claims like "It's okay to hit people just because you feel like it." Difference cultures believe different things, and it is not absolutely true or false that it's okay to hit people just because you feel like it.

Forty-three percent of participants gave the non-objectivist answer (answer #3).[1] Nichols found this general pattern of non-objectivist intuitions across a number of different scenarios. These results suggest that a sizable percentage of people appear to have non-objectivist intuitions about some canonical moral violation (see also Sarkissian, Park, Tien, Wright, and Knobe (2011)).

In Nichols's experiments, there was a substantial non-objectivist minority. Other research suggests that non-objectivists are more likely than objectivists to engage in creative problem solving when presented with a puzzle (Goodwin & Darley, 2006) and were more accepting of alternative viewpoints (J. C. Wright, Cullum, & Schwab, 2008). Creative problem solving and being more accepting of alternative viewpoints are tendencies that are typical of the personality trait *openness to experience*. Compared to others, people who are open to experience tend to be (a) more receptive to a variety of different experiences, (b) less likely to reason in accordance with accepted societal standards, and (c) more individualistic (John & Srivastava, 1999). It stands to reason, then, that those who are more open to experience may be more open to the possibility that there are no objectively true or false moral statements (Feltz & Cokely, 2008).

[1] Nichols also gave a non-moral case where two people disagreed about whether the earth was flat. Only 13% responded that there was no fact of the matter about whether the earth was flat. This was to control people who had more encompassing non-objectivist views. These people were excluded in Nichols's analyses. However, if they were included, 50% of participants gave non-objectivist responses.

The same basic strategy linking personality to philosophical relevant intuitions was used again here. Participants were given a description that is meant to capture the relevant aspects of moral objectivism. In this case, participants were undergraduates in a lower-level philosophy class recruited from a large state university. Those participants received Moral. Following Nichols (2004a), participants were asked to respond using one of the options 1–3 listed above. Those who responded (1) or (2) were operationalized as objectivists and those who responded (3) were operationalized as non-objectivists. Then participants were given the Ten Item Personality Inventory (Gosling et al., 2003). In this study, we were also interested in and wanted to statistically control other psychological factors that might be related to moral objectivism. So, participants completed (a) a Cognitive Reflection Task (the CRT) (Frederick, 2005) (b) a questionnaire about the number of philosophy classes completed, and (c) a self-report life satisfaction instrument (Diener, Emmons, Larsen, & Griffin, 1985).

The majority of participants (N = 79, 69%) gave the non-objectivist answer to Moral (non-objectivist scores (choosing option 3 above) were coded as 1; objectivist (choosing option 1 or 2 above) as 0). As predicted, those who were open to experience were more likely to respond as non-objectivists, $r(109) = .32$, $p = .001$. Judgments of moral objectivism were unrelated to all other personality traits (i.e., extraversion, conscientiousness, agreeableness, and emotional stability), sex, philosophical training, and reflective decision making ($p > .7$) (see Table 4.1).[2] Planned hierarchical

[2] If theories about the relation between life satisfaction and moral objectivism are right, they would primarily apply to those who were both conventional (i.e., lower in openness to experience) and non-objectivists. To simplify, the "natural" state of affairs for people with a conventional personality type (i.e., people who see things as clear-cut or "black and white") would tend toward objectivism (i.e., canonical moral issues are clear-cut, too). Thus, conventional people who don't see paradigmatic moral issues as clear-cut seem more likely to incur psychological costs including reduced satisfaction with life. However, correlational analysis indicated that moral objectivism was generally unrelated to satisfaction with life, ($p > .9$), as predicted. Analyses next examined the relations between openness to experience and life satisfaction for participants who expressed non-objectivist intuitions. An extreme group analysis of variance (ANOVA) was conducted examining top (open) and bottom (conventional) quartiles. An ANOVA with openness to experience (top quartile, bottom quartile) as a between-subjects factor revealed a difference in life satisfaction, $F(1, 36) = 6.61$, $p = .01$, $\eta_p^2 = .16$. Open non-objectivists were higher ($M = 26.9$, $SD = 5.55$) than average ($M = 25.1$, $SD = 5.85$) on ratings of subjective well-being. However, conventional (i.e., low openness) non-objectivists were about half a standard deviation lower than average on ratings of subjective well-being ($M = 22.1$, $SD = 5.87$).

Table 4.1 Intercorrelations for main variables. Females were coded as 0 and males as 1

		1.	*2.*	*3.*	*4.*	*5.*
1.	Moral objectivism					
2.	Openness to experience	.32**				
3.	Satisfaction with life	-.01	.32**			
4.	Reflective decision making	-.01	.01	.10		
5.	Philosophical training	.03	.08	.07	.19*	
6.	Sex	.03	.24*	.15	-.28**	.04

Notes: * $p < .05$; ** $p < .01$

regression analysis with all of the aforementioned variables predicted moral objectivism, $F\ (10, 100) = 1.96$, $p = .04$, $R^2 = .16$ (full model). Controlling for all other factors, openness to experience accounted for unique variance, $F\,(1, 100) = 9.64$, $p = .002$, $R^2_{\text{change}} = .08$.[3]

Given what we know about personality traits, it was probable that the relation between personality and non-objectivist intuitions could be predictably manipulated just as they were in intentional action and free will. We reasoned that we may be able to predict non-objectivist intuitions with a different personality trait by changing relevant aspects of the scenario. The action in Moral is supposed to be a canonical moral violation that involves an unjustified harm to another individual. However, there are other violations that do not involve harming anyone else. For example, there has been some research that some people are willing to say that disgusting actions that do not harm anyone else are morally wrong. A number of these types of actions are offered by Haidt, Koller, and Dias (1993). One action involves a man who buys a frozen chicken, takes it home, has sex with it, and then eats it. Arguably this action does not harm the man or anyone else (e.g., the chicken is already dead, the man does it in complete privacy). When Haidt et al. (1993) gave participants the Chicken scenario, about 65% of people responded that it would be OK if countries differed with respect to customs about having sex with dead defrosted chickens. However, nearly everyone in their study thought that having sex with a dead chicken was disgusting. This result suggests that many people

[3] In a nationally representative sample of people younger than 30, the relation between non-objectivist intuitions and openness to experience replicated, $r\ (37) = .29$, $p = .08$, see also Beebe and Sackris (2016).

think that there is no fact of the matter about whether it is OK to have sex with a dead chicken (although testing objectivist intuitions was not the main goal of Haidt et al.'s study).

Given the socially abnormal nature of having sex with a dead chicken, we hypothesized that those who were more socially minded would also be more bothered by the disgusting, bizarre action. As already discussed, extraverts tend to be more socially minded, tend to have relatively less emotional regulation, are more motivated to engage in social activities, and encode and recall socially relevant information differently. An action that essentially involves a socially abnormal, disgusting behavior seemed likely to exert a specific influence on extraverts as compared to non-extraverts. We predicted that extraverts would be more likely than introverts to think that there is a moral fact of the matter about having sex with a dead chicken.

We tested whether changing the nature of the action to a disgusting, yet harmless, act would alter the relation of personality to objectivist intuitions (Feltz & Cokely, 2008). In this case, all the participants were recruited from a psychology department's undergraduate student participant pool. Participants were presented with the following scenario that is a hybrid of scenarios used by Nichols (2004a) and Haidt et al. (1993):

> *Harmful chicken:* John and Fred are members of different cultures. They are in an argument about a newspaper article describing a man, Barney, who bought a frozen chicken, took it home, defrosted it, had sex with it, and then ate it. The article notes that doctors interviewed said there was nothing medically dangerous about having sex with and then eating the chicken (for example, salmonella is not transmitted via sex and the chicken was very well cooked). John says, "It's okay to have sex with a chicken and then eat it just because you feel like it," and Fred says, "No, it is not okay to have sex with a chicken and then eat it just because you feel like it." John then says, "Look you are wrong. Everyone I know agrees that it's okay to do that." Fred responds, "Oh no, you are the one who is mistaken. Everyone I know agrees that it's not okay to do that."

Participants were asked if the action was harmful. They were also asked if the action was wrong. As predicted, extraversion was related to harm judgments $r\ (145) = .24$, $p = .003$. Extraverts were also more likely to think that the action was wrong $r\ (146) = .23$, $p = .005$. However, when we looked at relations among the other four personality traits, no significant relation was found between those personality traits and harm or wrongness judgments.

Taken together, these studies provide two examples showing that two different personality traits may be predictably related to meta-ethical intuitions. While these are just some of the possible meta-ethical intuitions that one could have, these results suggest that at a minimum some meta-ethical intuitions can be predicted by some heritable personality traits. But meta-ethical intuitions are just one class of ethically relevant intuitions. Next, we turn to intuitions about first-order ethics.

Personality Predicts Bias in First-Order Ethics

First-order ethics is a prominent and distinct area of ethics. Rather than focusing on questions *about* ethics as meta-ethics does (e.g., moral objectivism), first-order ethics attempts to provide ethical principles or theories that help evaluate morally right or wrong actions. Largely, two general views about first-order ethics have occupied theorists. These two general approaches to ethics are consequentialism and deontology. Consequentialism is the view that the right making feature of an action is that the action creates the most good out of all alternatives (although what exactly constitutes the "good" is a contested notion; McNaughton & Rawling, 2006). Deontology is the other traditionally dominant view. According to deontology, the right making feature of an action is whether the action satisfies the correct set of principles or duties. Determining what the correct set of duties or principles is can be complicated and context-sensitive, but one critical difference between deontology and consequentialism is that some of the principles for the deontologist may not maximize the good (e.g., Ross (1988)).

Virtue ethics, an ancient approach to ethics emphasizing moral character, is a third approach to ethics that has made a resurgence since the last half of the twentieth century (Driver, 2001). One reason why virtue ethics has become popular again is that some theorists think that virtues can explain and inspire parts of our moral experience that are difficult for and traditionally neglected by consequentialists and deontologists. For example, consequentialism and deontology (or at least their simple versions) may not give adequate weight to many of the important internal dispositions of an agent (however, see (Copp & Sobel, 2004; Hursthouse, 1999; Slote, 2001; Swanton, 2003)). For example, acting courageously seems to be a morally important feature of actions in some situations. If consequentialists want to take into account the moral relevance of acting courageously, they must do so by explaining how courage promotes the good. In other words, they have to treat courage as a non-basic good. This feature of consequentialism appears to be contrary to everyday moral thought. It seems that sometimes one can behave courageously and do the right thing even if that behavior does not maximize the good. Similar complaints can be levied against the rule orientated nature of deontology. Virtue ethics shifts the focus of moral evaluation from maximizing the good or following rules to the motivations and character of a person.[4]

While virtue theories are diverse, there are some shared common themes (see Oakley (1996) for a fuller discussion of how to characterize virtue ethics and how it is different from consequentialism and deontology). For many, "the focus is on the virtuous individual and on those inner traits, dispositions, and motives that qualify her as being virtuous" (Slote, 2001, p. 4). "It is widely agreed that virtue is a trait of character" (Copp & Sobel, 2004, p. 516). In turn, "virtue is the concept of something that makes its possessor good" (Hursthouse, 1999, p. 13). Along these lines, Hursthouse holds that, "an action is right [if and only if] it is what a virtuous agent would characteristically (i.e., acting in character) do in the circumstances" (1999, p. 28). Many of these features that have been identified as right making features of actions go beyond the correct set of moral rules or producing the best consequences and involve deep seated dispositions to act and sets of motivational states. The following definition of virtue will suffice for our purposes:

> A virtue is a good quality of character, more specifically a disposition to respond to, or acknowledge, items within its field or fields in an excellent or good enough way. (Swanton, 2003, p. 19)

[4]Another way to see the distinction between virtue ethics and the other two dominant approaches to ethics is by reflecting on the *basis* for moral evaluation (Oakley, 1996). Swanson has a similar view:

> A virtuous agent has a standing commitment to act from virtue. This contrasts with that of the 'sophisticated consequentialist' agent who, according to Railton, 'is someone who has a standing commitment to leading an objectively consequentialist life'.... It also differs from sophisticated Kantianism, which demands that the Kantian moral agent has a standing commitment to perform her duty. (2003, pp. 28–29)

Virtue is often characterized to be or be a consequence of a character trait (Appiah, 2008; Aristotle, Ross, & Urmson, 1980; Calder, 2007; Driver, 2001; Hurka, 2006, 2010; Hursthouse, 1999; Langton, 2001; Slote, 2001). Even if some virtue theorists do not think that virtue is a character trait, most hold that virtue is a disposition to act (Hooker, 2002). These character traits or dispositions can be displayed in many ways (i.e., dispositions to respond) toward many different things (i.e., field of virtue). For example, one could be compassionate to a grieving friend by offering kind words or by offering friendly hugs. The grieving friend would be in compassion's field, and the kind words or friendly hugs would be the response. One need not be maximally compassionate to be virtuous. Rather, one must respond with a sufficient amount of compassion to be virtuous.

As will come as no surprise at this point, virtue ethicists take everyday intuitions seriously. These intuitions are often about particular cases. Indeed, a perusal of the literature reveals a number of specific references to cases and "our" intuitions about them (Slote, 2001) (Driver, 2001) (Hursthouse, 1999), (Copp & Sobel, 2004). Just as was the case in free will and intentional action judgments, these cases are designed so that the reader has an intuition in response to them. And, just as was the case in free will and intentional action, the intuitions about these cases are often thought to be widespread and shared by non-ethicists. To illustrate, Rosalind Hursthouse, a prominent virtue ethicist, writes that when "we" have intuitive reactions to cases, she means the "we" "to mean 'me and you, my readers'" (Hursthouse, 1999, p. 8). Michael Slote, also a prominent virtue ethicist, concurs and takes everyday intuitions seriously. As he write, "intuitive considerations...have considerable weight" (Slote, 2001, p. 5). In this light, many virtue ethicists desire that their theories accord with everyday thought about virtues and vices.

In the next three sections, we review some evidence about everyday thought about virtues and how that evidence can inform some debates in virtue ethics. Some people find that actions done from virtuous motivations are morally better than actions done from duty or to maximize the good, that consequences of character traits are a major factor in identifying character traits as virtues, and that sometimes epistemic imperfections can be necessary for full expressions of virtue. Importantly, across all of the studies in the next three sections the global personality trait emotional stability was found to predict virtue attribution.

Virtue, Deontology, and Consequentialism

As already mentioned, one of the primary motivations for virtue ethics is a dissatisfaction with the way that consequentialism and deontology handle some apparent morally relevant factors such as motivation and character. Virtue theorists often hold that virtues and vices naturally account for this part of our moral experience since motivation and character take central roles in virtuous and vicious actions. Many virtue ethicists take the pervasiveness of the attitude that virtuous motivation matters above and beyond maximizing the good or following the correct moral rule as support for their general argumentative strategy. But how pervasive is the intuition that virtuous motivations uniquely matter above and beyond consequentialist or deontological motivations? For whom are these internal states of the agent likely to matter?

Consistent with the leitmotif of this book, consulting empirical science most efficiently helps answer whether and for whom virtuous motivations matter. Some recent experimental work suggests that at least in some instances, actions stemming from virtuous motivation are judged to be morally better than actions stemming from consequentialist or deontological motivations (Cokely & Feltz, 2011). In one experiment, people were given a description of two people. One person acted from what is described as the correct moral rule generating better consequences and the other person acted only from a virtuous disposition:

> *Virtue:* Imagine two people, John and William, work in a hospital. They both witness 10 medical errors and learn that the hospital will be investigated for every error that is reported. Each investigation requires that the hospital must close for one week. When the hospital is closed, needy patients will be turned away.
>
> Person 1: John makes moral decisions solely based on consequences and widely accepted moral rules. John thinks long and hard to help with his decision and he calculates that the best consequences and the right moral rule dictate that he should report only 1 out of the 10 medical errors. For these reasons, and only for these reasons, John reports 1 medical error.
>
> Person 2: William does not make moral decisions solely based on consequences or widely accepted moral rules. Rather, William has the deep-seated character traits of justice and honesty that cause him to decide to report all 10 errors. Because of these character traits, and only because of these character traits, he reports all 10 medical errors.

Participants then answered the following question: "Whose action is morally better?". Participants were then given a 7-point scale where the scale's anchors indicated a preference for John (Likert scale value = 1) or William (Likert scale value = 7). A response of 4 would indicate that the participant did not have a preference for John or William. The overall mean response was 5.29, $SD = 1.94$ indicating a reliable judgment that the action done from virtue is morally better, overall and on average (i.e., William's action) ($t\,(41) = 4.29$, $p < .001$, $d = 0.67$).

While the difference in ratings of John's and William's motivation may be interesting and may support some of the general claims that virtue ethicists make, that result is not our primary interest. Our primary interest is whether preferences for virtuous or other motivations could be predicted by a global personality trait. There are some empirical reasons to think that personality is related to some virtue-related attributions. For example, those who report that they are more virtuous on some paradigmatic virtues also tend to be higher in the personality trait emotional stability (Cawley, Martin, & Johnson, 2000; Peterson & Seligman, 2004). People who rate themselves as low on emotional stability tend to have a wider spectrum of emotional reactivity and experience. That is, it is not simply the case that those who are more emotionally unstable are necessarily close to having emotional breakdowns or mental health disorders. Rather, they may simply tend to experience wider, more intense emotional variability than those who are higher in emotional stability. Nevertheless, people who rate themselves as highly emotionally unstable individuals have been characterized as being especially tense, anxious, nervous, moody, worrying, touchy, and fearful compared to those who are higher in emotional stability (John & Srivastava, 1999). Since anxious, moody, worrying, and fearful mental states are typically not thought to be virtuous (everything else being equal), it stands to reason that those who attribute those states to themselves would likely be less likely to think that they have the internal mental states that would count as virtues. However, does emotional stability predict attribution of virtue and moral worth to *others*?

To help answer that question, the general strategy we have been using throughout this book was applied. Participants responded to Virtue and completed a brief measure of the Big Five personality traits (the Ten Item Personality Inventory) (Gosling et al., 2003). Consistent with the relation of emotional stability with self-reports of virtue, emotional stability was related to judgments of the moral worth of the action, $r\,(40) = .36$, $p = .02$, indicating a preference for the action done with virtuous motivation (i.e., William's action). No other personality traits were reliably related to judgments about the moral worth of the action (all *ps* > .10).[5]

Even though there were good a priori reasons for thinking that emotional stability would be related to preferences for actions done from virtue, replication was still desirable. A second experiment was conducted to verify the relation of emotional stability to intuitions about the moral worth of an agent's motivations. In the second experiment, Virtue was slightly modified so that John and William were described as presidents of two different hospitals. This change was made to ensure that participants were not responding to some idiosyncratic feature of the original scenario. The first sentence of Virtue was replaced with the following sentence: "Imagine two people, John and William, who are presidents of a hospital." The beginning of the first sentence of the second paragraph was replaced with "John is president of hospital X and..." and the beginning of the first sentence of the third paragraph was replaced with "William is president of hospital Y and..." Participants were asked about the moral worth of the action. Again, there was a preference for the action done from virtuous motivation $M = 5.23$, SD 2.05, $t(52) = 4.35$, $p < .001$. Emotional stability was again related to judgments of the moral worth of the action, $r(53) = .38$, $p = .004$. No other personality traits were reliably related to judgments of moral worth (*p*s > .17).

These two experiments suggest that acting from virtue, as opposed to acting to generate the best consequences or follow the right moral rule, can influence people's moral judgments and feelings about the moral worth of actions. Moreover, these judgments may be systematically related to the global personality trait emotional stability. Those who were emotionally stable had a consistent and stable preference for actions done from virtuous motivation compared to actions done from deontological or consequentialist reasons.

Virtue and Consequences

One recent debate in virtue ethics involves how to identify virtues and virtuous actions. Some think that the only relevant factors for determining if someone behaves virtuously are their internal motivational states. Call

[5] A pilot study revealed the same relation of moral judgments about virtuous motivation and emotional stability, $r(136) = .17$, $p = .04$.

this view *evaluational internalism*. Evaluational internalism is prominently defended by Michael Slote. For Slote:

> [A]n act is morally acceptable if and only if it comes from good or virtuous motivation involving benevolence or caring (about the well-being of others) or at least doesn't come from bad or inferior motives involving malice or indifference to humanity. The emphasis on motivation will then be fundamental if the theory claims that certain forms of overall motivation are, intuitively, morally good and approvable in themselves and apart from their consequences or the possibility of grounding them in certain rules or principles. (2001, p. 38)

One motivation that Slote thinks is involved in virtuous behavior is compassion. On the evaluational internalist line, compassion is valuable independent of any consequences that acting with that motivation might have. If we only judge the motivational states of the person, then that helps dramatically reduce problematic cases of moral luck. Moral luck can be characterized as the possibility that sometimes bad motivations can produce good things and good motivations can sometimes produce bad things through no fault of the person acting (e.g., it was just random, dumb luck that those consequences came about). To illustrate, someone with a truly compassionate attitude may donate money to charity that ends up funding a cruel warlord unbeknownst to the compassionate person. Or one may maliciously slash another person's car tires with the consequence of saving the car owner's life (perhaps by the car owner noticing the brakes are severely and dangerously damaged when the tire is repaired). If the consequences of internal motivational states are what determine virtues, then the former internal trait should be considered a vice and the latter a virtue. But that assessment runs contrary to what many of us think about those situations. At least some of us judge the donor as being virtuous and the car tire slasher as being vicious. Judgments such as these are core to the evaluational internalist case because they are argued to capture a prominent aspect of our moral cognition, namely, that the internal motivational structures are what matter to virtue attribution and not the consequences of those motivational structures (Slote, 2001).

Others think that the only factors relevant to determining virtue are external to one's motivational state. *Evaluational externalism*, defended prominently by Julia Driver (1995, 2001, 2004), holds that "the moral quality of a person's action or character is determined by factors external to agency" (Driver, 2001, p. 68). For Driver (2001, 2004), these external factors are the actual (and not just expected) consequences that a character trait brings about. Driver identifies virtue as "a character trait that produces more good (in the actual world) than not systematically" (2001, p. 82). That these character traits bring about good effects systematically is important. *Sometimes* virtues can lead to bad consequences, they just cannot do so typically. The good consequences that are brought about systematically rule out most cases of moral luck. Of course, we can imagine cases of systematic moral luck, it's just that those cases would also be rare and unlikely in the world in which we actually inhabit. Hence, on Driver's externalist account, character traits or dispositions that do not typically bring about good consequences are not virtues and those that do are.

Many evaluational internalists and externalists place a heavy burden on folk intuitions (Crisp, 2010). According to these theorists, moral views are "judged partly in terms of how much ordinary thinking they preserve" (Slote, 2001, p. 13). Some think that evaluational internalism is more plausible because it "seems to have intuitive advantages over its more familiar utilitarian/consequentialist analogues" (Slote, 2001, p. 28) and it "is intuitively obvious and in need of no further moral grounding" (Slote, 2001, p. 39). But evaluational externalists say much the same thing. Driver holds that evaluational externalism is preferable because "it better captures some of our intuitions about hard moral cases" (2004, p. 72). When we "see that we have misjudged the consequences of a trait, we change our judgments of the trait's status as a virtue" (Driver, 2004, p. 84). As Driver notes, "These observations provide a great deal of intuitive support for a consequentialist theory of virtue" (2004, p. 84).

Evaluational internalism and externalism offer clearly contrasting, empirical predictions about folk intuitions about some paradigmatic cases of virtue attribution. These predictions have been put to empirical test suggesting that the consequences of internal traits are a major factor in determining whether those traits are virtues (Feltz & Cokely, 2013b). In one experiment, participants were asked to read only one of four different scenarios where the good and bad consequences of an action were systematically varied. However, the internal traits were intentionally left unspecified.

Pat: Scientists have recently become interested in Pat and have conducted an experiment about Pat. The scientists have found the following fact about

> Pat. Pat has a character trait that, when exercised, results in good things *100% / 51% / 10% / 0%* of the time. These good things include other people feeling good, people trusting Pat, and people feeling protected. When Pat's character trait is exercised, bad things come about *0% / 49% / 90% / 100%* of the time. These bad things include people feeling ashamed of themselves, people feeling humiliated, and people feeling unsafe.

Participants then rated their agreement with the following statement on a 6-point scale (1 = Strongly disagree, 6 = Strongly agree): "Pat's character trait is a virtue." Means and 95% confidence intervals are reported in Fig. 4.1.

As the data in Fig. 4.1 suggest, there was a strong, statistically significant linear relation in judgments. This linear relation would not be predicted by the evaluational internalist because the relation suggests that judgments are sensitive to the good that is brought about by the character trait. In contrast to the evaluational internalist's hypothesis, there was a clear linear relation between virtue attribution and the good consequences that the character traits brought about. The more good that was brought about, the more like a virtue people thought the character trait was.

These results may be of interest for adjudicating the evaluational internalist/externalist debate in virtue ethics. However, what was especially relevant to our goals was whether personality could predict virtue attributions. To do so, we first controlled for the percentage of time the character

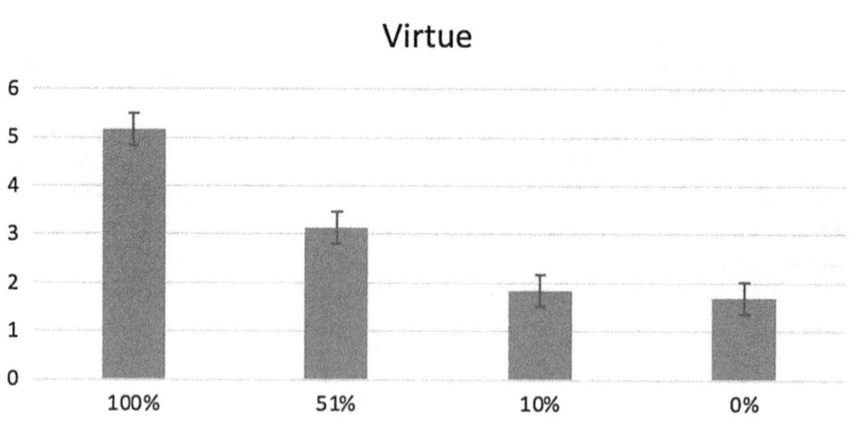

Fig. 4.1 Mean response to "Pat's character trait is a virtue." Error bars represent 95% confidence interval

trait brought about good consequences. Then we included all the Big Five personality traits in a regression model. The resulting regression indicated that extraversion and emotional stability predicted those who were likely to attribute virtue to Pat, F_{change} (2, 184) = 9.63, p = .004, $R^2{}_{change}$ = .03.

While the relation of emotional stability to attributions of virtue was expected, the relation of extraversion to these judgments was not predicted and required replication. Two follow-up studies were conducted to confirm and investigate why consequences mattered more than internal dispositional traits toward virtue attribution. In the first follow-up study, Pat was modified to control for a possible worry. In the description for Pat, we did not explicitly state that Pat was not aware of the good and bad consequences of Pat's traits. Because we did not stipulate that fact about Pat's mental states, it could be that at least some participants thought that Pat was aware of those good or bad consequences. That awareness is an internal mental state that could be relevant to evaluating Pat's character trait as a virtue. One way that might go is that if Pat had that knowledge and didn't change, that would reflect a non-virtuous motivation. And, the objection goes, that the non-virtuous motivation could be what is driving the linear effect found above and not the good or bad consequences of Pat's character trait.

To account for the potential effect of Pat's knowledge of the good or bad consequences, participants received slightly modified versions of Pat. In this version of Pat, we explicitly stated that Pat did not know about the consequence by including the following sentence at the end of the Pat scenario: "Through no fault of Pat, Pat is often unaware of these consequences."

Just as in the previous experiment, the results represented in Fig. 4.2 demonstrate a significant, strong linear relation as a function of the good or bad consequences of Pat's character trait. More importantly for our purposes, this experiment replicated the previously estimated relation between personality and virtue attributions. Using the exact same analytic strategy used above, we again found that extraversion and emotional stability predicted attributions of virtue to Pat F_{change} (2, 235) = 3.45, p = .03, $R^2{}_{change}$ = .01.

Even given the revised version of Pat, the evaluational internalist has a response that could explain the data while still being consistent with the pattern of results that would be expected from the evaluational internalist perspective. In either of the first two versions of Pat, it is left unspecified what motivations Pat has—that part is left completely unspecified. Given

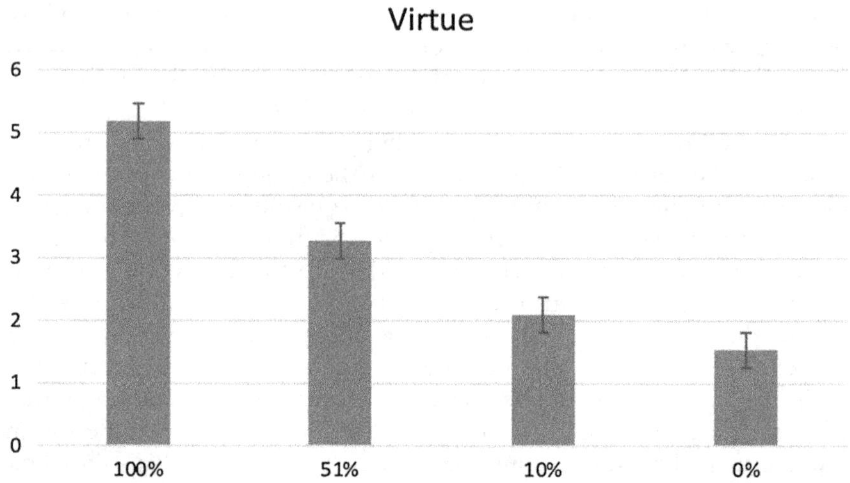

Fig. 4.2 Mean responses to "Pat's character trait is a virtue." Error bars represent the 95% confidence interval

the lack of specification of motivation, it could be that at least some participants are making inferences about Pat's motivation from the good or bad consequences. On this line of reasoning, if one produces good consequences, it is natural to think that the good consequences were produced by good motivations. However, when bad consequences come about, it may be natural to think that those bad consequences were produced by bad motivations. If at least some participants import this information about motivation into the study, then that could explain the linear effects while at the same time being consistent with evaluational internalism.

To address that worry, Pat was again revised. In this revision, the motivation with which Pat acted was specified. We created two different motivations for Pat's action. One group of participants read that Pat acted with a "desire to help others." Helping others is typically thought to be a virtuous motivation. A separate group of participants read that Pat acted with a motivation that was "indifferent to others." Indifference to others is taken by some to be a majorly defective and vicious motivation (Slote (2001)). One other change was made to the scenario to help head off objections: Pat was described as being "never aware of" the consequences of the trait (the previous modification of Pat only stated that Pat was *often* not aware). Given these changes, participants received one of the following versions of Pat.

> *Good/Bad motivation*
>
> Scientists have recently become interested in Pat and have conducted an experiment about Pat. The scientists have found the following fact about Pat. Pat has a character trait that makes Pat *desire to help/indifferent to* others. When exercised, this character trait results in good things *100% / 51% / 10% / 0%* of the time. These good things include other people feeling good, people trusting Pat, and people feeling protected. When Pat's character trait is exercised, bad things come about *0% / 49% / 90% / 100%* of the time. These bad things include people feeling ashamed of themselves, people feeling humiliated, and people feeling unsafe. Through no fault of Pat, Pat is never aware of these consequences.

For both Good and Bad motivation, the data were not consistent with what would be predicted by the evaluational internalist (see Fig. 4.3). However, the central question that concerned us was whether personality could predict virtue attributions. In the Good Motivation condition, a hierarchical linear regression model indicated that emotional stability was a predictor of virtue judgments when controlling for percentage of consequences, F_{change} (2, 218) = 4.78, p = .03, R^2_{change} = .01. A similar linear relation was present in Bad Motivation. A hierarchical linear regression model indicated that emotional stability was a predictor of virtue judgments in Bad Motivation when controlling for percentage of consequences, F_{change} (2, 190) = 3.99, p = .05, R^2_{change} = .02. However, in neither condition was the relation between virtue attribution and extraversion found to be reliable ($F_{change} < 1$). Consequently, this experiment replicated the relation of emotional stability but failed to replicate the relation of extraversion with judgments of virtue.

Overall, the data reported in this section suggest that character traits that bring about better consequences are, in some theoretically interesting instances, more likely to be thought of as virtues. In all three experiments, the global personality trait emotional stability predicted who was likely to attribute virtues. These results reinforce the results from the previous section indicating that some important types of moral intuitions about virtue are predicted by some global personality traits.

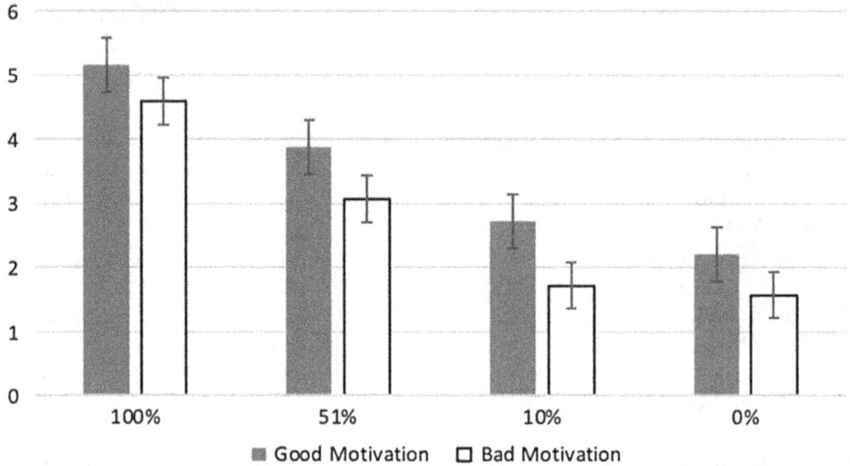

Fig. 4.3 Mean response to "Pat's character trait is a virtue." Error bars represent 95% confidence interval

Virtue and Ignorance

It is commonly thought that ignorance is bad, all else being equal. That is, ignorance illustrates some kind of intellectual deficit that could be corrected. Of course, sometimes ignorance is excusable (e.g., the authors of this book are ignorant of how to build a jet engine, but we also don't think we are particularly blameworthy for that ignorance). But sometimes ignorance is not excusable. Indeed, some ethicists have thought that the presence of ignorance is an epistemic vice and sometimes those epistemic vices can influence moral virtues. For example, Aristotle notes that "it is not possible to be good in the strict sense without practical wisdom" (1980, pp. 1144b, 1130–1131). Hursthouse claims that "each of the virtues involves getting things right" (1999, p. 12). She goes on to claim that "the agent must know what she is doing" (1999, p. 124). Swanton holds that "moral virtue has at its core rational virtue" (2010, p. 152). Brady states that it is a "deeply-held intuition...that the virtuous person (or at least fully virtuous person) counts as having both theoretical and practical wisdom, and this involves their having knowledge of what is the case and what is of value" (2005, pp. 93–94). Schueler claims it is "hard to see how it [virtue involving ignorance] differs from stupidity or self-deception, traits which may occasionally be useful but which are not usually thought of as virtues" (1997, p. 469). Smilansky argues that "ignorance and self-deception are not a good basis for virtue" (1997, p. 106). As we discussed at the beginning of this paragraph, some think that sometimes epistemic vices make actions less virtuous and the character traits less a virtue compared to when that epistemic vice was not present. For shorthand, we'll call this general view about the relation of epistemic vices and moral virtues *intellectualism*.

According to Driver (2001), intellectualism is wrong. She thinks that it is not a requirement of virtues that they involve true beliefs (e.g., a person has an epistemic defect that does not always result in a less perfect expression of a virtue). In fact, in some instances, ignorance is actually required for full expression of a virtue (Driver, 2003). Driver calls these kinds of virtues the *virtues of ignorance*.

Driver provides two ways epistemic defects could be relevant to virtue attribution. The first way is by a person having *propositional ignorance*. Propositional ignorance occurs when the person performing the action has some belief that is not true and is relevant to the action (Driver, 2001, p. 347). According to Driver, modesty is one example of a virtue that involves having a false belief about one's own value. While there are many accounts of what modesty is (see below), Driver argues the correct view is what she calls the underestimation account which is the view that "the modest person underestimates his self-worth to some limited degree" (Driver, 2001, p. 18). An underestimation constitutes a false belief (e.g., when you estimate that the jar of marbles at the fair has 500 marbles but actually has 625, your estimation is wrong). And since the false belief is about one's one value, it is relevant to the virtue of modesty.

The second way that a person could have an epistemic defect relevant to virtue is by engaging in incorrect inferences (from possibly true beliefs). This kind of epistemic defect Driver calls *inferential ignorance* (Driver, 2001, p. 347). Psychology has demonstrated numerous ways that one could have all true relevant beliefs but still fail to make inferences based on those beliefs (e.g., inattentiveness, time pressure, emotional reactions). In some of those instances, the failure to make the relevant inference could be important for the expression of some virtues. As we will see below, perhaps a fireman who does not even consider the danger to himself may be viewed as more courageous than a fireman who does consider all of the dangers and acts anyway. There may be something about the potential cost-benefit analysis that the latter fireman may engage in that makes the action seem less courageous than the former fireman. Importantly, it's not necessarily the case that either fireman has a false belief (in which case that could potentially be an example of propositional ignorance). In fact, it is very likely that because they are firemen they both have only true beliefs about the risks and benefits of acting. But still, there is some *epistemic* defect that the former fireman has that the latter one lacks—and this epistemic defect is likely to be failure to make an inference relevant to the virtue. Driver thinks that neither propositional ignorance nor inferential ignorance necessarily rule out virtue attributions. And, in some cases, they could be required for full expression of a virtue. Since we called the view that epistemic virtues are positively related to moral virtues intellectualism, we'll call Driver's alternative view ***anti-intellectualism***.

The way that we have characterized intellectualism dictates that any epistemic defect relevant to the virtue makes any virtue less good or less perfect. To the extent that ordinary intuitions about cases are supposed to be sensitive to factors involved in intellectualism, the anti-intellectualist and intellectualist positions give competing predictions about how attributions of moral virtue should function as a result of relevant epistemic failures. (We think it is a good bet that given what some virtue ethicists say, everyday judgments about cases are supposed to matter to the correct view about intellectualism, see discussion of important everyday intuitions to virtue ethics above.) When epistemic failures exist, there should be measurably lower moral virtue attributions if intellectualism is correct. The anti-intellectualist would predict it is not always the case that epistemic failures result in lowered moral virtue attribution. And, sometimes, the presence of an epistemic defect is related to more prefect expression of a virtue. As such, we should be able to construct hypothetical (or perhaps even actual) cases that capture those key contrasts and ask people to make judgments about them.

Since Driver (2001) uses modesty as her prime example, so will we. Just to recap, her underestimation account of modesty requires that one has a false belief about one's own value (with an important caveat that the false belief cannot be drastically wrong). As mentioned above, there are other prominent accounts of modesty. We will give a sampling of those that all seem to involve having true beliefs about one's own worth. The first is the *False Modesty* account. On this account, one has a true belief about one's own value but does not express that true belief. Rather, one intentionally downplays one's own sense of worth. For example, Einstein might have thought he was the best physicist ever. Assume that claim is true. On the

False Modesty view, Einstein would have that true belief but would downplay it. Perhaps he would say that he was merely a very good physicist. The other alternative view of modesty we will briefly look at is the *Accurate Modesty* account (Flanagan, 1990; Richards, 1988). This account of modesty involves having true beliefs about one's own value but also contextualizes that value. Take the Einstein example again. Rather than saying he was merely a good physicist, he could appreciate his contribution and argue that he was among the best physicists of the twentieth century. Here, he may not believe that he was the best physicist ever. In the Accurate Modesty account, Einstein still has all true beliefs (e.g., if it is a fact that he was the best physicist ever then it is true that he was the best physicist of the twentieth century, and he believes the latter but not the former) but he expresses those beliefs in a way that is more accurate to what he actually believes than he would on the False Modesty account.

Now we are in a position to start to address one of our central goals. Does personality predict virtue attributions? Again, our goal was to link everyday intuitions about virtue to some global personality traits. So, the standard strategy was applied. Scenarios were created and then personality was assessed. In this case, scenarios were created to reflect the distinctions discussed above (Feltz & Cokely, 2012b).

Ignorant Modesty

Albert Einstein is often thought to have sincerely said many times, "I am a good physicist." It is universally accepted that if Einstein was not the best physicist ever, he was one of the best physicists and certainly the best of the twentieth Century. Hence, Einstein falsely believed that he was "a good physicist" when in fact he was one of the best physicists ever.

False Modesty

Albert Einstein is often thought to have insincerely said many times, "I am a good physicist." It is universally accepted that if Einstein was not the best physicist ever, he was one of the best physicists and certainly the best of the twentieth Century. Hence, Einstein really believed that he was "one of the best physicists of the twentieth Century" when in fact he was one of the best physicists ever.

Accurate Modesty

Albert Einstein is often thought to have sincerely said many times, "I am the best physicist of the twentieth Century." It is universally accepted that if Einstein was not the best physicist ever, he was one of the best physicists and certainly the best of the twentieth Century. Hence, Einstein really believed that he was "the best physicists of the twentieth Century" when in fact he was one of the best physicists ever.

After reading only one of these scenarios, participants were given three prompts to respond to. The response options were on a 6-point scale (1 = strongly disagree, 6 = strongly agree).

1. "When Einstein said 'I am a good physicist/I am the best physicist of the twentieth Century' he exhibited a virtue.
2. Einstein was modest.
3. Einstein was a morally good person."

After responding to these prompts, participants completed the Ten Item Personality Inventory (Gosling et al., 2003).

Responses to scenarios indicated support for anti-intellectualism. While we will forgo the statistical analyses of these results since they are not our main concern, Table 4.2 shows that the overall results were not what the intellectualist would predict (see Feltz and Cokely (2012b) for the full analyses). Einstein's epistemic defect did not result in a reduction modesty attribution. In fact, that epistemic defect increased virtue attributions to Einstein. Consistent with the Pat scenarios, emotional stability predicted virtue attribution. For Ignorant Modesty, emotional stability was related to Virtue $r\,(43) = .37$, $p = .01$ and Good Person $r\,(43) = .35$, $p = .02$ but not Modesty $r\,(43) = .19$, $p = .22$. Emotional stability was unrelated to judgments in Accurate Modesty (r's < .14, p's > .26). In False Modesty, emotional stability was only related to Virtue ($r\,(63) = .26$, $p = .04$) and was unrelated to the other dependent variables ($r < .2$, p's > .11).

Two follow-up studies were conducted to ensure that there are some virtues of ignorance and to replicate the relation of emotional stability to

Table 4.2 Means and standard deviations for ignorant, false, and accurate modesty

	Ignorant modesty (N = 45)	*False modesty (N = 63)*	*Accurate modesty (N = 69)*
Virtue	$M = 4.82$, $SD = 0.86$	$M = 3.89$, $SD = 1.31$	$M = 3.41$, $SD = 1.49$
Modest	$M = 4.62$, $SD = 1.15$	$M = 4.05$, $SD = 1.56$	$M = 2.45$, $SD = 1.36$
Good person	$M = 4.53$, $SD = 1.08$	$M = 4.25$, $SD = 1.03$	$M = 4.01$, $SD = 1.11$

attributions of the virtues of ignorance. The first follow-up study controlled for beliefs that people may have about the historical figure Einstein. Maybe people have an idea of who Einstein was as a person (perhaps from film or other depictions). Or, maybe people have beliefs that since the twentieth century was the best century for physics, being the best physicist in the twentieth century is actually being the best physicist ever. In those cases, perhaps Einstein actually had true beliefs but underplayed them. If that is right, then the alternative views to the virtues of ignorance can account for the pattern of results.

To control for those worries, we ran an additional study to help address the potential issues with propositional ignorance. The major change was to remove Einstein from the scenarios and have the scenarios about some fictitious character named John who plays darts. The follow-up study also had one additional change. We added another scenario that involved John being unaware of how good of a darts player he is. We added this additional scenario because lack of awareness could also be a kind of propositional ignorance. Rather than having a false belief, John would lack a belief that might be reasonable for him to have. With all that in mind, participants read the following scenarios:

Ignorant Modesty

John is often thought to have sincerely said many times, "I am a good darts player." It is universally accepted that if John was not the best darts player ever, he was one of the best darts players and certainly the best of his generation. Hence, John falsely believed that he was "a good darts player" when in fact he was one of the best darts players ever.

False Modesty

John is often thought to have insincerely said many times, "I am a good darts player." It is universally accepted that if John was not the best darts player ever, he was one of the best darts players and certainly the best of his generation. Hence, John really believed that he was "one of the best darts players of my generation" when in fact he was one of the best darts players ever.

Accurate Modesty

John is often thought to have sincerely said many times, "I am one of the best darts players of my generation." It is universally accepted that if John was not the best darts player ever, he was one of the best darts players and certainly the best of his generation. Hence, John really believed that he was "one of the best darts players of my generation" when in fact he was one of the best darts players ever.

Unaware Modesty
John is often thought to have sincerely said many times, "I am a good darts player." It is universally accepted that if John was not the best darts player ever, he was one of the best darts players and certainly the best of his generation. Hence, John was unaware that he was "one of the best darts players of his generation" when in fact he was one of the best darts players ever.

Participants rated their agreement with the following statements on a 6-point scale (1 = strongly disagree, 6 = strongly agree):

(i) "When John said 'I am a good darts player/I am one of the best darts players of my generation' he exhibited a virtue.
(ii) John was modest.
(iii) John was a morally good person."

Participants then completed the Ten Item Personality Inventory.

The results from the follow-up study largely replicated the results of the first experiment about virtues of ignorance (see Table 4.3). Again, we will skip the fairly detailed statistical analyses, but the pattern of results does not support intellectualism about some virtues. Emotional stability was not related to judgments in ignorant modest (*rs* -.1 to .18, *ps* > .42). However, this result was somewhat expected because of the relatively small sample size's ability to detect relations of this magnitude. Yet participants' responses to Unaware Modesty, which features a kind of propositional ignorance, had nearly statistically significant relations to Virtue r (63) = .24, p = .057, Morally Good r (63) = .23, p = .066, but not to Modest r (63) = .18, p = .16. Emotional stability was unrelated to False

Table 4.3 Means and standard deviations for John the dart player scenarios

	Ignorant (N = 22)	*Unaware (N = 63)*	*False (N = 33)*	*Accurate (N = 45)*
Virtue	M = 4.0, SD = 1.51	M = 3.92, SD = 1.25	M = 3.48, SD = 1.48	M = 3.78, SD = 1.53
Modest	M = 4.77, SD = 1.34	M = 4.06 SD = 1.27	M = 3.3, SD = 1.69	M = 2.54, SD = 1.38
Good person	M = 4.41, SD = 1.18	M = 4.03, SD = 1.14	M = 3.73, SD = 1.26	M = 3.72, SD = 1.03

Modesty (*rs* .12–.28, *ps* > .11). Somewhat unexpectedly, emotional stability was related to judgments in Accurate Modesty: Virtue r (47) = .32, p = .03, Modest r (47) = .27, p = .06, but not Morally Good r (47) = .05, p = .72.

We conducted another experiment to estimate the other kind of epistemic defect's impact on virtue attribution—inferential ignorance. As the fireman case above may illustrate, Driver takes some types of courage to be good candidates for inferential ignorance. In particular, she thinks that impulsive courage "seems to involve inferential ignorance alone. The impulsively courageous person possesses certain relevant facts of his situation, yet fails to put these facts together in order to reach the conscious conclusion that he himself is in danger" (2001, p. 33). Driver states "a good illustration of this sort of person is one who, perhaps, fears for the person trapped inside a burning building but does not fear for himself, since he fails to represent the danger to himself" (2001, p. 33). The reader can probably already guess at the predictions. The intellectualist will predict that impulsively courageous acts will be judged less virtuous than non-impulsively courageous acts. The anti-intellectualist will predict the opposite. Once again, the main issue for our purposes was whether emotional stability predicted attributions of virtue.

To test inferential ignorance in the case of courage, participants read these scenarios:

Accurate Inference

Pat is taking a walk one night and comes across a burning building. Pat hears a child screaming for help inside the building. Pat is worried about the child inside the building. There is a high risk that if Pat goes into the building, Pat would get burned. Pat considers this risk very carefully. Pat accurately estimates the danger of going into the building when deliberating whether to go in. As a result, Pat fears for his own well-being. Pat decides to run into the building to save the child.

No Inference

Pat is taking a walk one night and comes across a burning building. Pat hears a child screaming for help inside the building. Pat is worried about the child inside the building. There is a high risk that if Pat goes into the building, Pat would get burned. Pat does not consider this risk. Because Pat doesn't consider this risk, Pat inaccurately estimates the danger of going into the building when deliberating whether to go in. As a result, Pat does not fear for his own well-being. Pat decides to run into the building to save the child.

Participants answered the following three questions about each scenario on a 6-point Likert scale (1 = strongly disagree, 6 = strongly agree).

1. "When Pat ran into the building, he exhibited a virtue.
2. Pat was courageous.
3. Pat was a morally good person."

After responding to these prompts, participants completed the Ten Item Personality Inventory.

Here, the data were less clearly supportive of the anti-intellectualist's predictions. As can be seen in Table 4.4, the strongest virtue attributions were in cases of Accurate Inference. Of course, the virtue attributions in all of the other cases were on the "agreement" side of the scale so the anti-intellectualist could argue that inferential ignorance does not rule out that one could be acting virtuously (even if the virtues are less perfect). But, more important for our purposes, did personality predict judgments of virtue? As expected, emotional stability did not predict judgments for No Inference (rs = – .05 – .02, $ps > .66$). However, for the other three cases, the data suggest trends (i.e., close but non-statistically significant differences in the right direction) where emotional stability predicted virtue attributions: Accurate Reasoning: Virtue r (94) = .13, p = .23, Courage r (94) = .19, p = .07, Good Person r (94) = .19, p = .06.[6]

On the whole, the series of experiments involving virtue has one consistent message: emotional stability tends to be predicably linked to attributions of virtues. This finding held regardless of whether participants were asked to compare virtuous actions to actions done from some other

Table 4.4 Means and standard deviations for accurate and inaccurate reasoning

	Accurate inference	*No inference*
Virtue	M = 5.18, SD = 1.12	M = 4.97, SD = 1.18
Courage	M = 5.5, SD = 1.09	M = 5.05, SD = 1.36
Good person	M = 5.32, SD = 1.06	M = 5.23, SD = 1.06

[6] We also conducted a separate pilot study on these scenarios with largely the same results for virtue attributions: Virtue: r (31) = .29, p = .11, Courage r (31) = .33, p = .07, and Good Person r (31) = .39, p = .03.

reasons, whether one would be virtuous while being ignorant, or whether virtues were identified by consequences. In short, personality predicts some first-order ethical intuitions.

Personality Predicts Bias in Applied Ethics

So far, we have reviewed instances where personality predicts philosophical intuitions about meta-ethics and first-order ethics. The final area of ethics we will consider is applied ethics. Where meta-ethics addresses issues about ethics and first-order ethics attempts to establish substantive ethical theories, applied ethics attempts to apply the lessons from these other areas of ethics to everyday moral issues such as abortion, suicide, and euthanasia. Of note, we talk as if these domains are isolated and discrete. In actual philosophical practice that is not necessarily the case. Results from applied ethics can inform theoretical work in first-order and meta-ethics as well (e.g., understanding applied ethical issues about animals may have important impacts for the scope of normative theories and may inform some ethical views about what a person is).

Almost all non-professional philosophers have beliefs or attitudes about almost all contemporary applied ethical issues (e.g., capital punishment, famine relief, vegetarianism). To be sure, there may be some areas of applied ethics that are less well-known to the non-expert (e.g., bio-enhancement or unmanned aerial vehicles), but these tend to be the exception rather than the rule. Of course, there can be substantial disagreement about what applied ethics is (Beauchamp, 2003). However, one common goal of applied ethics is to deal with moral issues as they currently exist in the world, including how people think about and interact with those issues (Singer, 1993). Hence, understanding and incorporating those everyday attitudes are important for the applied ethicist. For example, take a rule utilitarian approach to applied ethics where the correct action is the one that conforms to the *rule* that maximizes utility. Everyday thought is one important consideration when thinking about the costs or benefits associated with a rule. Implementing a rule that nobody should eat meat has costs because most people currently think it permissible to eat meat. Changing that attitude is likely to be difficult and expensive. It may end up that implementing such a rule is the correct thing to do according to the rule utilitarian. Determining the balance of costs and benefits includes understanding everyday thought about applied issues such as the morality of eating meat. Therefore, everyday attitudes can be one important factor for determining the correct moral rule.

It seems safe to conclude, then, that understanding everyday attitudes and decisions about many applied ethical issues are often essential in applied ethics. One implication of this conclusion is that it can be important to know *how* people come to have those beliefs or how they make their judgments. Take again the example of a rule utilitarian prohibition on eating meat. If we know how people come to believe that eating meat is morally permissible, we may be able to take steps to more efficiently institute the prohibition. Or we may find that the nature of humans' attitudes and decisions about eating meat are just too difficult to alter to expect the prohibition against eating meat to be effective. Similarly, we may find that some people's attitudes about eating meat rest on false beliefs (S. Feltz & Feltz, 2019). We may then be able to inform their decisions more fully (see Chap. 5 for an in-depth discussion of how these results factor into informed decision making). Or we may find that people currently are not capable of a dramatic shift in attitudes or decisions, but this does not mean that it will always be unreasonable to enforce the prohibition against eating meat. We may be able to take incremental steps so that future generations have different attitudes or make different decisions (O. Flanagan, 1991; Hoang, Feltz, Offer-Westort, & Feltz, 2023). All of this reinforces the notion that often it will be advantageous to know what current moral beliefs and attitudes are, and it will also be advantageous to understand the factors influencing those beliefs and attitudes and what role these features play in ethical decision making.

The past half-century has seen an abundance of research suggesting that people often use heuristics when making decisions (Gigerenzer, Todd, & ABC Research Group, 1999; Daniel Kahneman, Slovic, & Tversky, 1982; Simon, 1955, 1990). Heuristics are quick, efficient, and typically robust decision making rules of thumb that operate particularly well under constraints often found in real-world human decision making (e.g., time constraints, resource constraints). For example, one common heuristic is the representativeness heuristic. We often make judgments that are informed by our stereotype or idea of the typical event or person. Often these judgments are accurate (e.g., making judgments in reference to what a typical driver would do at a stop sign). However, in other instances those judgments can be mistaken (e.g., making judgments about the typical "man").

Sunstein (2005) speculates that a number of heuristics are involved in some people's "real-world" moral decision making (see also Gigerenzer (2010)). Sunstein calls one of the heuristics the ***Punish, and not reward, betrayals of trust*** (Sunstein, 2005, p. 537). Generally, people tend to prefer products that might decrease risk of harm as long as no portion of that harm comes from the product itself. For example, people preferred an airbag where 2% of people died in serious accidents compared to an airbag where 1.01% of people died but 0.01% of that risk resulted from the airbag itself (Gershoff & Koehler, 2011; Koehler & Gershoff, 2003, 2005). Sunstein speculates that this aversion would translate into judgments of punishment. Those who produce safer products would be deemed a more apt target of punishment than the producer of the less safe product if part of the risk of harm came from the safer product itself. As such, this heuristic would lead to the odd result of punishing a company that produces an overall *safer* product.

But could personality predict who acts as if they use the punish and not reward betrayals of trust heuristic? To help answer this question, we present results from two new studies, following the same basic strategy typically employed to link personality with philosophically relevant judgments. We first gave participants a vignette that captures the philosophically relevant features of the ***Punish, and not reward, betrayals of trust*** heuristic and then we assessed people's personality.

In the first study, 62 participants were recruited from Amazon's Mechanical Turk. Participants received the following scenario. Nine participants were excluded for not completing the survey. Three participants were excluded for giving obviously inconsistent responses (e.g., expressing they would prefer to Buy Cure A but then expressing a stronger preference for Cure B). Thirty-five were male (70%). The mean age was 27.69, $SD = 7.78$ ranging from 18–53.

> *Deadly Virus:* Suppose that you are exposed to a deadly virus that kills 100% of those infected. Suppose that you are offered a choice between two equally priced cures: Cure A and Cure B. Scientific tests indicate that there is a 2% chance that those who take Cure A and are exposed to the virus will be killed due to the virus. Scientific tests indicate that there is a 1% chance that those who take Cure B and are exposed to virus will die due to the virus. However, Cure B may kill people who would not have died if they took Cure A instead. Specifically, some of those who take Cure B may die due to becoming especially vulnerable to the virus. Tests indicate that there is an additional one

chance in 10,000 (0.0001%) that someone who is exposed to the virus and takes vaccine B will be killed due to becoming especially vulnerable.

After reading this scenario, participants chose which of the two Cures they would prefer to buy (Cure A or Cure B). They also then responded to the following prompt: Please indicate the strength of your preference on the scale below (1 indicates a strong preference for Cure A. 7 indicates a very strong preference for Cure B). Participants then completed the Ten Item Personality Inventory and the Berlin Numeracy Test (BNT). Among educated adults living in industrialized countries, the BNT has been found to be the best single predictor of general decision making skill including one's ability to understand and evaluate risk (i.e., risk literacy) (Cokely et al., 2012, 2013, 2014; Ghazal et al. 2014). Of note, the numeracy test typically fully mediates the relations between general cognitive abilities (e.g., intelligence, attentional control, working memory) and superior judgment and decision making performance in non-expert samples (e.g., more intelligent people make better decisions because they have better numeracy skills, just as less intelligent people make better decisions because they have better numeracy skills). Often this test more than doubles the predictive power of other instruments (e.g., 10 times better than 30-minute tests of fluid intelligence). In other words, in general samples there is typically no test that will be more sensitive or likely to detect systematic relations between individual differences in general cognitive abilities and choice. If numeracy is unrelated, it's a very good bet that the influence of other general abilities is trivial (see *Skilled Decision Theory*; Cokely, Feltz, Ghazal, Allan, Petrova, & Garcia-Retamero, 2018).

In response to Deadly Virus, most people preferred to buy Cure A, $N = 33$ (66%), $\chi^2 = 5.12$, $p = .02$. This preference was also reflected in the strength of the preference for Cure A from neutrality ($M = 1.34$, $SD = 0.48$), $t\,(49) = 39.31$, $p < .001$. Conscientiousness was unrelated to the binary choice of which cure one would prefer to buy *rho* (50) = -.15, $p = .31$, but was related to a stronger preference for buying Cure A *rho* (50) = -.28, $p = .05$. BNT was related to the preference for which drug to buy *rho* (50) = .40, $p = .004$, and showed a trend predicting strength of the preference *rho* (50) = .23, $p = .11$. To determine the unique predictive power of BNT and conscientiousness, a multiple linear regression with conscientiousness and BNT as independent variables and binary cure choice as the dependent variable was conducted. The full model was a strong and significant predictor of the binary choice: $F\,(2, 47) = 6.24$, $p = .004$, $R^2 =$

.21. Results showed that BNT was a unique predictor of the binary choice ($\beta = .13$, $t\,(47) = 3.33$, $p = .002$) while conscientiousness showed a negative numerical shift where more conscientious individuals were less likely to take the safer vaccine ($\beta = -.04$, $t\,(47) = -1.43$, $p = .16$). BNT and conscientiousness did not interact ($t < 1$). A similar multiple regression was conducted with preference scale as the dependent variable. The full model was a significant predictor of the scaled choice: $F\,(2, 47) = 4.03$, $p = .02$, $R^2 = .15$. Results indicated that BNT was a marginally significant unique predictor ($\beta = .28$, $t\,(47) = 1.94$, $p = .059$) while conscientious was a robust unique predictor ($\beta = -.22$, $t\,(47) = -2.21$, $p = .03$) of the scaled choice. BNT and conscientiousness did not interact ($t < 1$). These results suggest that those who are more numerate prefer the overall safer product while more conscientious individuals preferred a less safe product where the product itself had no chance of killing them.

The relations between BNT, conscientiousness, and preferences for vaccines accord with Skilled Decision Theory (Cokely et al., 2018). Those who are more numerate are more likely to deliberate more carefully and elaborately leading to more informed and normatively correct responses in many domains. For example, numerate individuals are more likely to make correct choices in paradigmatic risky prospect interpretation tasks (e.g., evaluating the risks and payoffs of various financial gambles; selecting more effective medical treatments; making public policy recommendations that save more lives with fewer costs; making judgments about the risk of climate and weather hazards). Given that the goal of the vaccine is to reduce death, those who are more numerate may pick the vaccine that reduces overall risk of death. However, those who are more conscientious may be especially sensitive to the "betrayal" vaccine B represents. Those who are conscientious tend to hold that duties are important. If a duty of a vaccine is not to kill a person, then conscientious people may prefer a less safe product that has no chance of killing people.

While the pattern of results generally accords with broad theoretical frameworks, post hoc explanations are generally easier and less reliable compared to a priori prediction (e.g., it is hard to predict what the stock market will do tomorrow but relatively easy to "explain" what you saw happen yesterday). Once again, what is required for increased confidence in a result's robustness is replication. Thus, a follow-up study was conducted using slightly different materials. The structure of the scenario was identical, but instead of cures for a disease we used airbags with similar descriptions and identical probabilities of harm used in Deadly Virus

(Gershoff & Koehler, 2011; Koehler & Gershoff, 2003, 2005). Participants responded to the same questions except they were modified to reflect that the scenario was about airbags. Participants also received the Cognitive Reflection Task (CRT) instead of the BNT. As a reminder, CRT measures the degree to which one tends to express a more impulsive cognitive style, which is a predictor of numeracy and decision making more broadly (Frederick, 2008). Eighty-six undergraduate students were recruited from a large public university. Fifty (51%) were male. The mean age was 19.44, $SD = 1.76$ ranging from 18–28.

In this revised case, there were no differences in preference for the less safe airbag ($N = 44$, 51%) compared to the safer yet harm causing airbag ($N = 42$, 49%), $\chi^2 = 0.05$, $p = .83$. Preferences for the safer airbag measured by the Likert scale were also not significantly different from neutrality, $M = 4.08$, $SD = 1.45$, $t(85) = 0.52$, $p =.6$. However, conscientiousness was associated with a stronger preference for the less safe car airbag on the binary choice, *rho* (86) = =.31, $p = .004$, and with a stronger preference for the less safe airbag on the scaled response: *rho* (86) = -.2, $p = .066$. CRT was also related to the binary choice *rho* (86) = .22, $p = .04$ and was trending on the scaled response *rho* (86) = .2, $p = .069$. To determine if conscientiousness and CRT were unique predictors, a multiple regression with CRT and conscientiousness as independent variables and binary car choice as dependent variable was conducted. The full model was a significant predictor. $F(2, 83) = 6.37$, $p = .003$, $R^2 = .13$. CRT was again found to be a near significant predictor of choice ($\beta = .09$, $t\ (83) = 1.86$, $p = .066$) while conscientious was a clear and significant unique predictor ($\beta = -.06$, $t\ (83) = -2.56$, $p = .01$). A similar multiple regression analysis was performed for the scaled car choice. The full model was a significant predictor of the scaled response: $F(2, 83) = 3.69$, $p = .03$, $R^2 = .08$. CRT was a significant unique predictor ($\beta = .3$, $t\ (83) = 2$, $p = .05$) while conscientious showed the expected trend toward unique prediction ($\beta = -.09$, $t\ (83) = -1.36$, $p = .18$).

In summary, these results provide a demonstration of one way cognitive and personality variables may have independent and opposing influences on behavior. Specifically, conscientiousness (i.e., a heritable and stable trait) tends to be an independent predictor of the degree to which participants acted in accord with the ***punish, not reward, betrayals of trust*** heuristic. In contrast, the cognitive skill and style instruments predicted choices that were consistent with less reliance on the heuristic. Theoretically, the results may reflect differences in one's primary motivation or focus. If one focuses on the overall consequences (e.g., the numbers of harmed individuals), as may be likely for more numerate individuals, one may be less likely to act as if trust has been betrayed. However, when one instead focuses attention on the potential costs of betrayals of trust, it may be harder to ignore or consider other factors particularly among those who care deeply about moral commitment and obligation (i.e., "dutifulness" is a primary facet of conscientiousness). Taken together, the cognitive and personality factors uniquely contributed to a predictive model of choice, providing a more comprehensive and nuanced account of the judgment processes that give rise to these ethically relevant choices.

Empirical Science and the Autonomy of Ethics

We'd like to address one initial worry that one might have specifically about empirical science and ethics—a worry that does not necessarily exist for empirical science's relation to free will or intentional action. Ethics is a normative endeavor—it tells us how we ought to act or behave. Since ethics is a normative discipline, one might object that the actual way we are or think about ethics is irrelevant to what we ought to think or how we ought to act. One may concede that there may be some important role in applied ethics for how we actually think and behave, but that role would be only secondary. Only after we come to know what the right moral principles or actions are should we worry about how we actually are or how we actually go about making decisions. For these reasons, one might think that ethics is autonomous from empirical psychology and that empirical results can in principle have nothing to say about most substantive theoretical pursuits in ethics.

Owen Flanagan (1991) identifies two ways in which ethics could be thought to be autonomous from empirical findings. The first way is that ethics is a moral standard setter. On this view, the job of ethics is to discover the principles or rules about what is morally permissible, obligatory, or forbidden. Since ethics in this sense is a normative discipline, the way that people actually believe or behave can say nothing about how they ought to believe or behave. Flanagan notes parallel reasoning in epistemology. The way people actually reason may not be important to how they ought to reason. Epistemology sets the standards for how people ought to reason. As such, ethics, like epistemology, is autonomous from empirical psychology (even if it can help inform ethical ways to implement findings from empirical psychology, e.g., in moral education).

A second way that ethics could be autonomous from empirical psychology is that ethics deals with the analysis of moral concepts or of moral linguistic terms. The task of ethics is to analyze moral concepts or terms such as "right," "wrong," and "obligatory." The goal, according to Flanagan, is to find a common conceptual scheme that underwrites the usage of these terms, to systematize discrepancies, and to clarify ambiguities in these terms. This approach is a "semantic" analysis of those terms or concepts and as such is autonomous from empirical psychology since the goal is a philosophical conceptual analysis of the terms.

These objections have a long and celebrated history in ethics. There have been many defenses and responses to the claim that ethics is autonomous. Flanagan gives a particular useful analysis of the problem. On Flanagan's view, the autonomy thesis is too strong on both of these approaches. For the standard setting approach, the way humans are (or can be) provides a constraint on acceptable moral standards. After all, it is unreasonable to set a standard that no humans could possibly satisfy.[7] The results from ethics should be *psychologically realizable*.

Empirical psychology can help inform us of what kinds of creatures humans are and what is psychologically realizable for humans. To illustrate, virtue ethicists are perhaps the most prominent group of moral theorists who respect the psychological realizability constraint on ethics. As we have indicated above, it is common for virtue ethicists to think that how "we" (including non-professionals) judge cases and what intuitions we have are important for theory construction in virtue ethics (e.g., Anscombe, 1958; Appiah, 2008; Arpaly, 2003; Driver, 2001, 2004; Foot, 2001; Hursthouse, 1999; Zagzebski, 2010). One reason is that virtues are often thought to be something that humans can have since virtues are often characterized as being character traits (Doris, 2002; Harman, 1999; C. Miller, 2013). As a corollary, humans should also be able to evaluate those character traits in terms of how virtuous or vicious they are. Given these observations, it seems like a significant tension for the virtue theorist if those judgments or actual character traits were completely irrelevant to

[7] It is tricky to spell out the possibility claim here. Flanagan thinks that a reasonable moral standard that no present human could meet can still be reasonable to the extent that the standard is one that it is at least possible that some (future) human could meet. To help elucidate this notion, Flanagan introduces what he calls the degree of difficulty. To the extent that a standard requires a greater modification to human psychology in order to be instantiated, that standard has a higher degree of difficulty. Standards can have high degrees of difficulty to instantiate and not be impossible.

the correct account of virtue (e.g., akin to saying a virtuous person is a river—if true, virtue is not a theory of humans because humans are not and never could literally be rivers). So, any principled rejection of empirical psychology is saddled not only with the challenge of psychological realizability, it is also saddled with rejecting a tradition where those judgments and empirical evidence are thought to be valuable. And, as we argue in the next chapter, rejection of tradition is not to be done lightly or without good evidence. Of course, this does not mean that empirical psychology settles debates. All that is claimed is that empirical psychology is relevant to ethics and that ethics is not completely autonomous from the empirical evidence. Philosophical reflection still has important roles to play even if not fully autonomous.

The linguistic analysis approach straightforwardly does not support the strong autonomy thesis. Flanagan first notes that ethics only as linguistic analysis is so far removed from essential notions of what ethics is and is supposed to be that such an approach cannot be taken seriously. For example, it is highly unlikely that non-professional philosophers think that ethics aims to analyze ethical terms or concepts. Rather, ethics, among other things, is meant to aid in practical reasoning. Ethics is supposed to tell us what we ought to do and why. Moreover, since the analysis is of everyday concepts and usage, there is little reason to think that philosophers are uniquely positioned to do that kind of research. Rather, psychology and allied behavioral sciences have the tools and methods to efficiently address how everyday terms or concepts are used and when. In other words, "Ethics conceived as semantic analysis is just a kind of descriptive social psychology—possibly poorly executed" (O. Flanagan, 1991, p. 28).[8]

All of this is consistent with some weak autonomy thesis. One need not think that ethics is reducible to empirical psychology. There are some important roles for philosophical reflection to play. First, setting standards and ideals that are not currently realized can be one important role that ethics can play. For example, virtue ethicists do not always take the

[8] Flanagan notes that ethics could be autonomous from empirical psychology if ethics is religiously non-naturalistic. These non-naturalistic views are diverse and difficult to neatly characterize. On many of these views, however, human psychology is important. For example, on Judeo-Christian conceptions, it is important that humans be good. It would be odd if God created us so that we are the kinds of creatures that could not be good. On others, human psychology may not be important because it is impossible for humans to live according to God's standards and living well is only obtained by luck. These latter views are the exception on Flanagan's view.

empirical evidence without question or treat empirical evidence as the only kind of evidence. Second, data may need a certain degree of systematization and refinement in light of other plausible principles (Zagzebski, 2010). Data don't interpret themselves, and philosophical reflection can help with that interpretation. Importantly, these views still respect the psychological realizability constraint. As such, we follow Flanagan in saying "all moral theories—certainly all modern ones—make our motivational structure, our personality possibilities, relevant in setting their moral sights. The autonomy thesis, the claim that psychology does not matter to moral philosophy, can be safely ignored" (1991, p. 31).

Conclusion

In this chapter, we have reviewed some empirical results suggesting that several heritable personality traits predict ethically relevant intuitions in meta-ethics, first-order ethics, and applied ethics. Overall, we have provided a detailed and representative (albeit non-comprehensive) account of the most relevant, currently available evidence. We have also discussed an important objection to using everyday intuitions and empirical science. This objection holds that ethics as a normative discipline is completely autonomous from empirical science. The objection has been found wanting. Any ethical theory that does not respect the psychological realizability principle is implausible and impractical. In fact, most contemporary ethical theories respect and attempt to account for the empirical evidence. We take it, therefore, that empirical evidence is relevant to most philosophical pursuits—even ones that are normative and that have traditionally been argued to be completely independent of empirical results. While giving sustained arguments for this claim is outside the scope of this chapter, others have convincingly made this point (O. J. Flanagan, 1991). There are other objections and concerns remaining to be answered, but Chaps. 1–4 set the stage for the rest of the book.

The major focus of the last three chapters was on intuitions about hypothetical and actual cases. However, it is valuable to again note that personality also predicts philosophically and morally relevant behaviors. There is no room or need for an exhaustive review of the many philosophically relevant behaviors that personality predicts because the evidence and breadth of these connections is very extensive and well-established (e.g., Ashton et al., 2012). To illustrate, personality traits are so robustly related to occupational performance, workplace citizenship, theft, absenteeism, tardiness, lack of cooperation, and other forms of counterproductive and illegal behavior that personality assessment is a common and legal practice used in the hiring of employees in the United States and elsewhere.

In case the "real-world" association between personality and philosophical (e.g., ethical) behavior is not yet compelling, in closing it seems useful to consider some more extreme variations in personality, namely *personality disorders*. Personality disorders are diagnosed and defined as personality tendencies that are enduring, deviant, distressful, dysfunctional, disruptive, and not the result of some kind of extenuating circumstances (e.g., prolonged drug use) (Association, 2013). Given that these disorders are dysfunctional, disruptive, and distressful, it is not surprisingly that some personality disorders are associated with some morally undesirable and even morally reprehensible behavior. For example, there has been extensive research conducted on *Antisocial Personality Disorder*. Diagnostic criteria include tendencies that are abnormal, interfere with good interpersonal and personal functioning, and are persistent (Cooke, Forth, & Hare, 1998). People with Antisocial Personality Disorder who exhibit pronounced antisocial behaviors tend to be impulsive risk takers who have relatively low levels of empathy and muted emotional reactions. Not surprisingly, these individuals have higher rates of criminal activity and recidivism (Leistico, Salekin, DeCoster, & Rogers, 2008; Patrick, 2006). In some extreme cases, for most of us it is nearly unimaginable how little remorse some such individuals show for appalling, vicious, and disgusting crimes (e.g., torture of children, cannibalism, and necrophilia). Thankfully, such extreme cases are rare, as are personality disorders generally. Nevertheless, taken together with other evidence from typical individual, we think these examples should help make it clear that stable, heritable dispositions (i.e., personality traits) are often related to individual differences in philosophically and morally relevant thoughts and behaviors. Overall, we take it that we have established a theoretically and empirically informed case that personality often predicts fundamental philosophical intuitions. We now turn to some major implications these findings have and some emerging opportunities across frontiers.

CHAPTER 5

Philosophical Expertise

Imagine that your dog didn't want to eat his food and he was lethargic. He wasn't interested in going for a walk and didn't even want his favorite toy. He normally is very hungry and loves to play, so you know something is wrong with your dog, but you don't know exactly what. Who would you go to find out what is wrong with your dog? Why would you go to see that person?

It seems likely you would go see your veterinarian and for good reasons. The veterinarian has trained for years to be able to know what to look for, has developed skills to be able to make good decisions, often rather quickly, and has the practice to be able to use that knowledge. Of course, veterinarians are not perfect—they sometimes make mistakes. But on average, they are able to display consistent, repeatable, and superior performance in taking care of your dog compared to the average non-veterinarian. In other words, your veterinarian has *expertise* and can display *expert performance* concerning your dog's health (Ericsson et al., 2006, 2007, 2018; Cokely et al., 2018). This expertise likely makes the veterinarian's medical judgment about your dog better than your or your friends' judgments. It also means that your veterinarian's judgment is less likely to be influenced by factors that are extraneous to correct diagnoses and treatment (e.g., the veterinarian's personality or mood). In this way, through training and deliberative practice, your veterinarian's judgments about your dog's health are likely to consistently just be *better* than non-veterinarian's judgments.

A. Feltz, E. T. Cokely, *Diversity and Disagreement*,
https://doi.org/10.1007/978-3-031-61935-9_5

One might think that something similar happens when one becomes a philosophical expert. On this line of thought, philosophical experts, through their training and deliberative practice, have better philosophically relevant intuitions than non-philosophical experts. Philosophers may have a better grasp of the key concepts and arguments or they may simply have more apt cognitive styles and strategies that make them more tuned to key elements of scenarios. In turn, the knowledge or cognitive styles may make their intuitions better, truer, and less susceptible to mistakes. This reasoning holds that philosophers, just like veterinarians, are therefore less likely to be influenced by extraneous factors like personality concerning their judgments in their area of expertise.

The reasoning expressed in the previous paragraph is consistent with what has been come to be known as the *Expertise* Defense. The Expertise Defense holds that through philosophers' special training and abilities, extraneous features are less likely to influence their expert judgments. In this chapter, we review the Expertise Defense. We then go on to criticize the Expertise Defense in two ways. First, we provide arguments that the kind of expertise philosophers are likely to have is not likely to make some of their philosophical judgments better or less prone to problematic biases (for more on what exactly is problematic about these biases, see Chap. 6). Second, we provide direct evidence that at least some extraneous factors influence philosophers' judgments in some paradigmatic examples. These two criticisms suggest that the Expertise Defense fails, at least as it pertains to philosophers' reliance on intuitions for some central philosophical projects in some prominent philosophical areas.

The Expertise Defense

Perhaps the most common response to the empirical data about potentially problematic philosophical implications of variation in philosophically relevant intuitions is the Expertise Defense. The Expertise Defense has been articulated in various forms by several theorists (Sosa 2007a; Kauppinen, 2007; Ludwig 2007; Williamson, 2007, 2011). However, a common theme that unites all the various forms of the Expertise Defenses is the basic notion that philosophers are different from non-philosophers in one very important way. Unlike non-philosophers, philosophers are experts about the area in which they work. Like most fields, philosophy has theories and terms that are nuanced and difficult to understand. Philosophers, compared to the folk, are likely to understand these nuances and theories better. That richer and more sophisticated understanding of those terms and theories would make it less likely that those who are experts would be prone to the same problematic biases or judgment tendencies as the folk. Of course, these expert philosophers can incorporate folk intuitions into their theorizing, including potentially problematic biases, but the philosophers themselves are not likely to display those same problematic biases. If the Expertise Defense is correct, then the kinds of associations of philosophically relevant intuitions and personality we've documented in Chaps. 2–4 may be interesting but are not that problematic for the practice of philosophy.

We think that we have made the case that for many philosophical projects, intuitions play important (perhaps irreplaceable) roles. As such, these general approaches constitute an important element in the tradition of philosophy. We take tradition seriously. While there are often good reasons to alter or reject tradition, we take the position that there needs to be good reason to change or alter tradition. In other words, we accept that the burden of proof rests on us (see Horvarth, 2010; Sosa, 2009; T. Williamson, 2011). That means we need to provide reasons why the philosophical tradition of giving a central role to philosophy needs to be altered or abandoned.

At this point, a few words about what a burden of proof is and what we view as a reasonable assessment of when that burden has been satisfied are warranted. First, we take it as a truism that the burden of proof can be satisfied. Some theorists appear to set an exceptionally high bar. For example, Kauppinen observes that "the actual studies conducted so far have failed to rule out competence failures, performance failures, and the potential influence of pragmatic factors" (2007, p. 105). We agree. But empirical science simply cannot ever satisfy that burden—it is always *possible* that some other factor is responsible for observed effects. Empirical science just is not in the business of ruling out everything. Consequently, the burden has to be something less than that.

Some have attempted to satisfy this burden in one way or another (see, for some examples, below). Concerning the Expertise Defense, we adopt the position that the burden of proof has been satisfied when we demonstrate that expert philosophical intuitions vary similarly (or similarly problematically) with folk intuitions (J. M. Weinberg, Gonnerman, Buckner, & Alexander, 2010, p. 333). There is no need to satisfy the stronger claim that experts have exactly the same intuitions as the folks and are biased in exactly the same ways. After all, and consistent with our view presented below, expertise does matter in lots of domains and in many ways. And, there is no need to claim that personality alone is responsible for variation in some philosophical relevant intuitions (a very implausible view). Rather, we think that all we need is evidence that personality is among the factors that are responsible for philosophically relevant intuitions in expert philosophers. As such, one of the goals in this chapter is to provide evidence that expert philosophical intuitions are related to personality. If we achieve that goal, then we think we have largely discharged the burden of proof set out by defenders of intuitions in many philosophical projects.

Before we evaluate the Expertise Defense, we'll start by offering a somewhat lengthy discussion concerning the scientific evidence about (a) what expertise is, (b) how people can acquire expertise, and (c) how we measure expertise. These three elements are essential to understanding what the Expertise Defense amounts to and whether we can tell if philosophers have the relevant expertise to deflect worries about their intuitions. Those not interested in an in-depth discussion of expertise and how it is developed can safely skip to the summary of key points on page 181.

Decision Making

One useful framework in psychology holds that human judgment and decision making performance is often characterized by the interplay of "fast and slow" thinking processes, also often referred to as differences in Dual Systems (Kahneman, 2011). The idea is that "humans have, in effect, two separate minds" (Evans & Frankish, 2009). System 1 is said to be evolutionarily older and rapidly gives rise to intuitions and emotion. The other system, System 2, is evolutionarily newer and thought to be primarily involved in deliberative and coherent rational thought. More specifically, the evolutionarily older *System 1* (i.e., "fast") processes may typically involve high capacity, fast, associative, parallel, unconscious, and automatic processes that give rise to intuitions and impressions. In contrast, *System 2* ("slow") processes are more evolutionarily unique to humans and generally tend to be slower, effortful, serial (i.e., unfolding one step at a time), and conscious, involving rule-based processes that are demanding of working memory and executive functions (e.g., activity in the dorsolateral prefrontal cortex of the brain). For the current purposes, and in accord with most available data, the Systems are often characterized as having a default-interventionalist architecture: System 1 generates intuitions based on past experience, associations, and emotions, while System 2 then monitors and potentially corrects or modifies those intuitions with logic or deliberation, assuming sufficient attentional resources and motivation (e.g., when one is not stressed, checked out, or thinking about too many things; see Kahneman, 2003, 2011).

The Dual Systems approach has been widely adopted, connecting research in most subfields of psychology as well as neuroscience, economics, and philosophy (Kahneman, 2003; Stanovich & West, 2000). The evidence that human cognition can be efficiently characterized by differences in automatic (e.g., intuitive) and deliberative processes is well-established and has been for about four decades (e.g., automatic v. controlled processes; Shiffrin & Schneider, 1977). Of course, there are some serious concerns about the specific instantiations of the dual systems theory and its predictive validity (Cokely, 2009; Gigerenzer & Regier, 1996; Moshman, 2000; Newstead, 2000; Newell, 1973; Osman, 2004). Nevertheless, the framework has proven very popular and useful in the decision sciences because of its broad explanatory power: Even though the theory does not allow for many specific predictions, it does help organize and interpret a wide range of results.

Dual Systems theory has been put to extensive use to explain the link between domain-general cognitive abilities and normatively superior (i.e., high-quality) decision making. To be clear, domain-general cognitive abilities refer to abilities, like one's attentional control (e.g., working-memory capacity; Cokely, Kelley, & Gilchrist, 2006), that tend to be at least a little beneficial on lots of tasks (e.g., a person who is better able to regulate their attention can do so on tasks at work and on tasks at home). In contrast domain-specific abilities like expertise tend to be profoundly beneficial for very specific tasks only (e.g., a chess master is excellent at chess but will be no better than any other amateur when presented with a different strategy game like Poker). Research shows that domain-general cognitive abilities, including intelligence, statistical numeracy, working memory, attentional control, and others, tend to predict more normative judgment and decision making in classical heuristics and biases tasks (i.e., abstract, laboratory experiences like choosing between risky gambles; see Cokely & Kelley, 2009; Cokely, Feltz, Ghazal Allan, Petrova, Garcia-Retamero, 2018).[1]

[1] We invite the interested reader to visit www.RiskLiteracy.org for a brief automated test of their numeracy skill, one known to predict a wide range of beneficial risky decision making tasks (e.g., interpreting medical risks, understanding information economic data).

In theory, the relationship between general abilities and superior decision making reflects differences in the interplay of System 1 and System 2 processes. More intelligent people are more likely to use System 2 to monitor and correct the output of System 1, or else they may disregard biased intuitions all together and use normative rule-based processes to calculate answers. For example, when faced with a risky prospect (e.g., the choice between two gambles), individuals with higher levels of attentional control tend to make more correct choices. That is, participants tend to act *as-if* they weight and integrate the available information in accord with an expected value model (i.e., multiply value by probability and select the option that will on average offer the highest expected payoff). Interestingly, however, research shows that even in highly simplified, paradigmatic tasks "smarter" people don't tend to use more normative processes to make better judgments and decisions. Instead, System 2 processes appear to reflect *qualitative and quantitative* differences in simple deliberative heuristic search and problem understanding (i.e., elaborative encoding of stimuli) (Cokely & Kelley, 2009; see also Barton et al., 2009; Woller-Carter et al., 2012; and for examples in expertise, see Moxley, Ericsson). Theoretically, the links between domain-general abilities and rational decisions reflect a host of *early selection* metacognitive processes (i.e., thinking about thinking; Cokely & Kelley, 2009; Flavell, 1979).

What Is a Good Decision?

The emergence of the modern scientific debate on human rationality, or how people make decisions and what qualifies as a good decision, can be traced in large part to the Ages of Reason and Enlightenment (i.e. seventeenth and eighteenth centuries, respectively). During these times logic and careful, justifiable reasoning became highly prized by philosophers, empiricists, and political actors alike. As an example, consider the astronomer and physicist Pierre-Simon Laplace. Laplace's legacy includes seminal contributions to probability theory; however, more important for our purposes, he also provided a description of a fictional omniscient being that captured the *Zeitgeist* of the times. This being, known as Laplace's super-intelligence, was envisioned as one who would know all the details of past and present and with this knowledge could readily make good choices and predict the future with perfect certainty (Gigerenzer, 2006).

For many people, Laplace's vision of a decision maker who is omniscient and computationally unbounded may seem like an elaborate fantasy.

Yet this fantasy or some version of it is fundamental to much of the research and theory in the modern economic, cognitive, and decision sciences. Some readers will find this surprising, or ironically unreasonable, but models of "rational man" and *homo economicus* are among the most central and influential models used in the allied decision and risk sciences. According to neo-classical economic theory people behave *as-if* they are *unboundedly rational* and make optimal (but not necessarily perfect) choices—choosing *as-if* they have solved a complicated decision calculus (Hastie, 2001; Shafir & Tversky, 1995). These decisions can be described by optimization processes that reflect people's maximization of their own subjective expected utilities (i.e., personal values) via multi-attribute integration calculations wherein one optimally weights and integrates all available information in the light of one's values and other risks or uncertainties. As a simplification for illustration, one could list every possible pro and con for a certain decision, weight each pro/con according to values (subjective utility), multiply those values by the probability of occurrence, and then integrate the information optimally (e.g., linear integration as in linear regression). Such theories are at the core of dozens of models of decision making including modern theories on diverse topics in motivation, attitudes, and moral judgments (Gigerenzer & Selten, 2001; Gigerenzer et al., 1999) (Weirich, 2004). Although this approach has provided interesting and useful theory, these models often conflict with empirical evidence as psychological science has clearly demonstrated that even though people act as-if they perform a complicated decision calculus, this is not how real people with limited resources (e.g., time, attention, memory) use information to make decisions (Gigerenzer, Todd, and the ABC Research Group, 1999; Kahneman, Slovic, & Tversky, 1982; Payne, Bettman, & Johnson, 1992, 1993; Shafir & Tversky, 1995). Indeed, many decisions are so computationally complex or underspecified that optimization would not be possible for any person or known machine (Gigerenzer et al., 1999).

In the mid-twentieth century, Herbert Simon (1955, 1990) introduced his notion of bounded rationality. Simon argued that people have only limited time, knowledge, and cognitive resources and thus human decision makers cannot carry out the types of optimization computations that were (and still are) often assumed to be essential to rational decision making. Instead, Simon argued that effective decision making must often involve heuristics, which can be less formally described as *simple rules of thumb* (i.e., non-optimizing decision processes with non-exhaustive search processes; Simon, 1990; for a computationally precise modern extension of this program, see Gigerenzer et al., 1999). In the 1970s, Daniel Kahneman and Amos Tversky carried related ideas forward with the acclaimed *heuristics and biases* research program (Kahneman, et al., 1982; Kahneman & Tversky, 2000; Tversky, & Kahneman, 1974). This research program provided a huge body of evidence showing that people often relied on a handful of heuristics that led to biases. Note that a bias is technically defined as a tendency—e.g., most people have a right-hand bias for most activities like writing—it is not necessarily synonymous with error but can be associated with errors under specific conditions. Interestingly, however, the heuristics and biases program focused extensively on biases that led to non-normative errors. While this provided vivid and illustrative examples, it also led to some confusion because identifying normative errors required normative assumptions and justifications about the appropriate standards for an accurate or good judgment, something that is not without controversy (Anderson, 1991; Gigerenzer, & Goldstein, 1996).

In the case of the heuristics and bias approach, it was assumed that human cognition should be compared to a very specific set of context-free normative standards such as the outcomes of "rational" optimization processes and logic. Thus, non-normative errors are said to be evidenced when people's judgments deviate from "an established fact…[or] an accepted rule of arithmetic, logic, or statistics" (Kahneman & Tversky, 1982, p. 493). For example, when asked "which city is further North: New York or Rome?" many people confidently respond *New York*, even though it is incorrect. Similarly, when a doctor says "95% of patients who are treated survive" people tend to feel much more optimistic about surgical outcomes than when a doctor says "5% of patients who are treated don't survive." Theoretically, the same information is provided yet differences in *framing* produce dramatic differences in intuitions and biases—a non-normative, non-logical difference. To further illustrate with one of the most influential findings to emerge from the heuristics and biases research program, consider the model of how people value risky prospects—i.e., prospect theory. A technical description is beyond our current scope, but a key component is reflected in the fact that people tend to prefer receiving $100 for certain when compared to a 75% chance of winning $200, yet paradoxically they prefer a 75% chance of losing $200 to a certain $100 loss. Theoretically, when faced with multiple lotteries such as these, the normative decision is one that simply calculates the expected value of the two prospects by multiplying the probability by the potential outcome and comparing the choices. Thus, we are comparing two prospects, one worth \$150 on average (i.e., 75% of \$200 = \$150) and another worth \$100 on average. Accordingly, it is rational, on average, to prefer the risky option (75% of \$200) for gains but not losses, even though most people prefer the exact opposite. To simplify, people act *as-if* losses loom larger than equivalent gains: The subjective joy one receives by gaining \$100 pales in comparison to the subjective pain of losing an equivalent amount, hence the pattern of risk aversion for gains and risk preference for losses (i.e., losses hurt almost three times more than the joy one experiences from an equivalent gain).

The impact of the research on heuristics and biases is hard to overestimate, having influenced our fundamental understanding of human psychology and behavior across many domains including medicine, finance, business, economics, and law. Nevertheless, despite its many successes, the heuristics and biases program has some notable limitations. One of the most serious concerns is that the program has emphasized ways in which heuristics are associated with errors, which has led some to argue that heuristic use is a problem that needs to be corrected. In this light, heuristics are seen as inferior or second-best choice processes designed to be used by computationally disadvantaged individuals. In contrast, research demonstrates that heuristics are often powerful tools that can lead to superior decision making in humans, animals, and machines, particularly under conditions of high complexity or uncertainty as are present in many everyday decisions (Gigerenzer 2008; Simon, 1990). Other concerns focus on the fact that when more representative materials are provided, many biases go away (e.g., Gigernezer, 2001). Still other work has emphasized important differences in criteria used to evaluate judgment and decision making, including coherence (e.g., logic and calculation) versus correspondence (e.g., predictive validity in natural environments). That is, some violations of neo-classic notions of rationality also appear to result from strategies that are very well adapted to real-world task requirements (McKenzie, 2003; Hammond, 2000). Setting this issue of the appropriate standard to the side, what is more central to our current review is the nature and interplay of intuitive and deliberative cognitive processes that are thought to give rise to more and less rational judgments and decisions.

To further illustrate, consider an example. In manufacturing one can improve the quality of goods sent to market by (a) improving inputs (e.g., more skilled workforce), (b) improving outputs (e.g., careful inspection and repair), or (c) doing both. In the metacognition literature these quality control efforts are referred to in terms of (a) early selection versus (b) late correction processing (Jacoby, Kelley, & McElree, 1999; Jacoby, Shimizu, Daniels, & Rhodes, 2005). Late correction processes attempt to detect and repair (e.g., System 2) the output of faulty automatic processes (e.g., System 1) such as biased intuitions. In contrast, early selection can use controlled processing (e.g., System 2) to generate goals, strategies, and mental contexts that qualitatively alter the output of automatic processes (e.g., System 1) before biased intuitions are generated. Research suggests that early selection processes may be key factors that influence a wide range of behaviors, including performance on intelligence tests themselves. Individuals who score higher on domain-general cognitive ability measures tend to spend more time preparing for tasks and also more elaborately and strategically encode information, deliberatively building cognitive representations that provide better support during subsequent task performance (Baron, 1978, 1985; Cokely & Kelley, 2009; Ericsson & Kintsch, 1995). To the extent that early selection cognitive control processes are recruited, they involve deliberate memory encoding. This elaborative encoding causes information in working memory to be moved to long-term memory, freeing-up attentional resources and creating more enduring and detailed mnemonic representations (Cokely, Kelley, & Gilchrist, 2006). In laboratory tasks, this tends to cause better task performance because better representations give rise to better intuitions and to a better ability to monitor performance. However, these same types of metacognitive and deliberative efforts are also processes that give rise, over time, to domain-specific expertise. Indeed, the Knowledge is Power account of Skilled Decision Theory suggests that the primary reason that most people make better decisions is not because they override intuitions but instead because they educate them (i.e., using System 2 to educate and refine System 1 so that intuitions are naturally more informed and less biased; Cokely et al., 2018; see also Cho et al., 2024).

Acquiring Expertise

Sometimes people find it surprising but today there is wide agreement among the scientists who have studied expertise that *experts are always made, never born*: Without exception, no matter how talented someone may be, they will need to practice deliberatively for many years before they will be able to become a verifiable expert performer. Research also shows that standardized general ability tests and genetic markers consistently fail to predict individual differences in expert performance, such that there is no correlation between intelligence and skilled performance in fields such as chess, music, sports, and medicine. Among typical and healthy individuals, two of the only innate differences that have been found reliable to predict success are height and weight, which matter to relatively few professions. So what does predict success? To put it simply, deliberative practice and access to valuable resources.

A considerable body of research has been devoted to studying the acquisition of expert performance, including the mechanisms that give rise to expert performance more generally (for a practically comprehensive treatment of relevant findings, we refer the interested reader to the Cambridge Handbook of Expertise and Expert Performance; Ericsson et al., 2006). Among the core findings of this research is that expertise doesn't just require practice, rather it requires a particular kind of practice known as *deliberate practice*, which primarily involves specific and sustained efforts at doing something one *couldn't* do before. Research also shows expertise requires a great amount of deliberate practice. All expert performers, including the most gifted or talented, need a minimum of about ten years (or 10,000 hours) of intense training before they succeed at the highest levels such as winning international competitions, which is an important criterion for high levels of expert performance. Of course, in some fields the apprenticeship is even longer. The most elite musicians often require on the average of 20–30 years of steady practice in order to succeed at the international level. The development of verifiable expert performance also requires specific kinds of environments. For example, Bloom's (1985) landmark study suggests that elite performers often study with devoted teachers and tend to be supported enthusiastically by their family and relatives throughout their developing years. More than this, however, experts need to be in a learning environment that is not systematically biased, and they need accurate and timely feedback on their performance. Without feedback one cannot learn. And if one does not learn, one never improves the quality of one's intuitions.

Several recent landmark studies have transformed our understanding of the causes and consequences of *general decision making skill.* Beyond informing contemporary theory and policy, these findings speak to longstanding debates about the association between general intelligence and decision making. For more than a century major assumptions about the nature of this link have shaped debates about the causes and consequences of class structure, which appear to have affected opportunity in many ways. A theoretical question at the heart of these debates is whether decision making ability is primarily determined by the basic, innate cognitive abilities of normal healthy people. At one extreme we know the answer: Expert performers who engage in extensive deliberate practice are consistently able to circumvent limitations imposed by basic, innate cognitive abilities thanks to long-term working-memory resources that support superior decision making within their domain of expertise (Ericsson & Kintsch, 1995; Ericsson, Prietula, & Cokely, 2007). But still, a persistent refrain is that innate, general cognitive abilities (e.g., intelligence) are important predictors of better decisions and in some instances set upper bounds of what a person can achieve cognitively.

To further illustrate this perspective, consider data from a related seminal contribution entitled *Cognitive reflection and decision making* (Frederick, 2005). In this research, Frederick showed that participants with higher general cognitive ability scores tended to act *as-if* they avoided fundamental biases similar to the framing effects previously discussed by weighting and integrating the available information in accord with an expected value model (i.e., multiplying the value by probability). All participants answered tricky yet rudimentary math-type questions that often bias people toward incorrect intuitively appealing answers (e.g., "if a bat and a ball cost 1.10, and the bat cost 1.00 more than the ball, how much does the ball cost"—hint it's not 10 cents). Those who answered these kinds of questions incorrectly tended to show marked asymmetries in their evaluation and selection of risky prospects (e.g., risk seeking for losses yet risk averse for gains). Although Frederick was cautious with his theoretical interpretations, converging evidence indicates that his assessment of cognitively impulsivity also predicted steeper rates of delay discounting on intertemporal choices, which was compelling for many reasons (e.g., which would you prefer $300 dollars now or $400 dollars next month).

Likewise, the groundbreaking work of Stanovich and West (1998, 2000, 2008; see also Toplak, West & Stanovich, 2011), Frederick (2005; Kahneman & Frederick, 2007), and others tells the story of the state-of-the-science at that time. The emerging leading perspective from the relative new field of decision psychology was that intelligent people generally tended to make more intelligent decisions because they possessed the special capacities needed to override non-rational, emotional, or intuitive impressions in support of complex logical and formal decision analyses. In some sense this was an efficient hypothesis to start with because among other virtues it was a simple explanation in accordance with economic assumptions. Nevertheless, it largely appears to be wrong.

Deliberation Is for Understanding

Building on the work of Peters, Baron, Reyna, and many others, Cokely and Kelley (2009) conducted the first study to directly map the relations between decision strategies, basic cognitive abilities, and superior decision making under risk. Using choice outcome modeling, decision latencies (e.g., reaction time), and retrospective verbal protocol analysis (Ericsson & Simon, 1984; Fox, Ericsson & Best, 2011) they assessed and modeled *how* individuals with higher cognitive ability scores (i.e., working memory, numeracy, and cognitive reflection) typically made superior decisions when evaluating paradigmatic risky prospects (i.e., lotteries). Despite the paradoxical findings from Peters et al. (2006) indicating that sometimes more skilled decision makers were more biased, dual systems theory suggested that more cognitively able individuals might generally make better decisions under risk by inhibiting affective responses and generating abstract and logical decision analyses (e.g., Evans & Frankish, 2009; Kahneman & Tversky, 2000; Kahneman, 2003). However, retrospective protocol analyses, wherein participants recreated their decision strategies after their decisions had been made, indicated that less than 5% of the sample attempted to calculate expected values during decision making. Instead, the vast majority of people made superior risky decisions because they tended to deliberate more, such that the ability-to-performance relationship was fully mediated by large differences in affective and elaborative evaluation and understanding (i.e., representing relations between feelings, thoughts, and consequences in personally relevant narratives).

The results of Cokely and Kelley (2009) showed that even when evaluating very simple risky prospects, superior decision making under risk generally followed from differences in how and how much participants thought about and understood the decision problem (see also Pachur & Galesic, 2013). For example, better decision makers spent more time imagining how changes in wealth would affect their life and how those changes might feel via informal narratives (e.g., "even though that's probably never going to happen it really is more money than I pay in tuition so I can't take the risk"). Generally, better decisions also appeared to reflect *metacognitive heuristics* that offer simple strategies for understanding and exploring thoughts and feelings, such as *disconfirming* (e.g., identifying multiple reasons for and against a decision), *reframing* (e.g., considering potential outcomes framed in terms of costs as well as potential benefits), *forecasting* (e.g., more elaborately exploring how potential consequences would feel for various stakeholders and why), ***prioritizing*** (e.g., reflecting on their own assumptions about what their goal was, why it was their goal, and what their top priority goals should be), and *re-checking* (e.g., transforming probabilities, re-reading, and organizing facts and assumptions).[2] Moreover, deliberation as measured either by number of considerations or decision latency predicted superior decision making much better than any (and every) other combination of cognitive ability test scores. Among these relatively typical public college students, decision making quality was much better explained by deliberative heuristics strategies and representative understanding than by cognitive ability profiles (e.g., cognitive impulsivity, attentional control) or logical formal decision analyses (e.g., expect utility). Indeed, some of the least "able" individuals were nevertheless among the best decision makers, reflecting their extensive and personally meaningful heuristic deliberation.

Theoretically, the relations between decision making skill and deliberative heuristic search reflect a host of metacognitive processes that are essential for ***understanding*** and contextualizing the decision problem (Cokely & Kelley, 2009; Cokely et al., 2012; Ghazal et al. 2014, Garcia-Retamero & Cokely, 2013a, 2013b, 2013c; see also Peters et al., 2006; Peters, 2012; Reyna et al., 2009; Reyna, 2004). This is useful in part because a more representative understanding of risks and trade-offs means that decision heuristics are better informed (e.g., accurate assessment of cue validities and the magnitude of stakes). The same way extensive knowledge and practice allow expert performers to quickly make superior decisions in routine situations by considering only a small number of cues (Shanteau, 1988, 1992), heuristic deliberation helps people identify the most essential information and trade-offs that take priority for heuristic decision making. Essentially, elaborative deliberation serves as a means of contextualizing risks and consequences in personally meaningful terms, which helps people intuitively feel the weight of various options and stakes without expressly creating or solving a formal econometric analyses.

Theoretically, the processes that support general decision making skill (and risk literacy) are the same as those that give rise to complex situation model development during reading comprehension and those that

[2] Related research has since revealed that training these kinds of metacognitive and elaborative heuristic reasoning strategies can substantially improve judgments and decisions that are relevant to geopolitical risks and hazards, such as those typically made by government intelligence analysts (Chang & Tetlock, 2016; Chang et al., 2016).

maintain the high-fidelity situation awareness that often characterizes expert performance. The common thread is that skilled decision making isn't usually limited by basic cognitive abilities because the development of an integrated understanding engages long-term working-memory capacities (Ericsson & Kintsch, 1995). In turn, long-term working memory functionally expands one's reasoning capacity far beyond what could be supported by basic cognitive capacities alone (e.g., attentional control). In effect, because decision makers have a vast expert-like knowledge of themselves (e.g., experiences, values, and preferences), personally meaningful heuristic deliberation enables fast and durable long-term memory encoding and representation of complex constellations of relevant risks, rewards, and trade-offs. That said, even if nearly anyone can functionally circumvent limitations imposed by basic cognitive abilities like intelligence by utilizing their ultra-high-capacity long-term working-memory resources, accurately evaluating risk still requires that people have specialized risk literacy skills, which is a topic we'll save for another time (e.g., for recent reviews, see Cokely et al., 2012, 2013, 2014, 2018; see also Allan et al., 2017a; 2017b; Barton et al., 2009; Cho et al., 2024; Ellis et al., 2014; Garcia-Retamero & Cokely, 2011, 2012, 2013a, 2013b, 2013c, 2014a, 2014b, 2015a, 2015b, 2017; Garcia-Retamero et al., 2012, 2014, 2015, 2016a, 2016b, 2019a, 2019b; Garrido et al., 2021; Ghazal et al., 2014; Keller et al. 2010; Merritt et al., 2010; Okan et al., 2012a, 2012b, 2015, 2018; Petrova et al., 2015, 2017, 2018, 2023; Petushek et al., 2014, 2015a, 2015b; Ghazal et al., 2014; Ramasubramanian et al., 2019; Raza et al., 2019, 2023; Salehi, et al., 2018; Wong et al., 2010; Woller-Carter et al., 2012; Ybarra et al., 2017).

Based on the literature reviewed in the past three sections, there are three points that are most relevant to assessing whether philosophers have expertise and what kind of expertise that is likely to be.

1. *Experts are always made and not born.* The principle applies to philosophical expertise as well. Nobody is born a philosophical expert. Acquisition of philosophical expertise requires prolonged, deliberate practice.
2. *Innate cognitive abilities* (e.g., *intelligence) can't explain and are not necessary for expert performance.* Again, this principle applies to philosophers. If there is philosophical expertise, being a philosophical expert does not require or entail that one possesses a rare level of

general cognitive abilities (e.g., intelligence) that non-experts are unlikely to possess.

3. *Expert judgment and decision making primarily result from differences in knowledge and skills, which enable long-term working memory to support fast, durable, and complex mental representations and processes.* Again, this principle applies to philosophers. If philosophical expertise exists, it will primarily reflect the acquisition of specialized skills and knowledge that will allow philosophers to use long-term working memory (instead of relying on limited short-term memory resources) to conceptualize, reason, and think in highly sophisticated ways.

Given these three principles, we can help answer the following questions: Is there philosophical expertise, and if there is, what is it and does it provide support for the Expertise Defense? In the next two sections, we review some evidence relevant to assessing philosophical expertise characterized by points 1–3.

Expertise in Philosophy

Indirect Strategies

Broadly speaking, there are two general strategies that one could use to determine the strength of the Expertise Defense—indirect and direct strategies. The first we will discuss are indirect strategies. Indirect strategies generally attempt to identify some of the key elements identifying expertise or how one can develop expertise. Then, those who adopt an indirect strategy try to argue that the ways that the markers of expertise or the ways that expertise are developed in philosophy do not have some of those key elements. Given the lack of some of these key elements of expertise, we would not expect that philosophers have the relevant kinds of expertise to deflect the worries raised by some results from experimental philosophy. In those cases where there is a lack of expertise, we would not expect that philosophers would have any qualitatively better intuitions (e.g., better early selection of intuitions or better late correction of intuitions).

J. M. Weinberg et al. (2010) have adopted indirect strategies to argue against the expertise defense that largely mirrors how we have documented expertise identification and development above. To illustrate one way expertise development in philosophy differs from other areas, we'll look at one of their examples—i.e., the kind of feedback that is provided. As reviewed above, feedback is one required element for developing expertise. For example, when piloting an airplane, it is clear when one makes a mistake. One goes off course, crashes an airplane, violates flight plans, etc. These mistakes are often not only evident but they also are not temporally remote from the action, both of which are important for being able to recognize and learn from the feedback that is given. But in many philosophical domains and debates, that kind of unequivocal evidence and immediate feedback is lacking. Except for some cases in logic or issues concerning factual knowledge (e.g., historical dates, specific examples, specific vocabulary or cases), our perspective is that it is rare for a philosophical view to be seen as simply mistaken by all. For example, is the Justified True Belief account of knowledge wrong? Do arguments with "mistakes" that involve the Justified True Belief account of knowledge have the same kinds of feedback mechanisms as those for the airplane pilot? It appears that, for the most part, the answer to both questions is "no." If that is right, then the proponent of the indirect strategy argues that philosophers are simply not likely to have the right kinds of feedback to make their intuitions qualitatively better and immune from effects documented in experimental philosophy (including personality's relation to some philosophically relevant intuitions).[3] Hence, philosophy doesn't provide the right kind of feedback to develop the relevant expertise. Since the relevant feedback is often lacking, expertise is not likely to be developed. So, if there is philosophical expertise, it is not of the relevant kind to help support the expertise defense.

While indirect strategies are suggestive that philosophy might not provide the right kind of feedback to develop the relevant philosophical expertise, it would be desirable to have some evidence that philosophers lack the relevant expertise to support the Expertise Defense. In other words, it would be desirable to have actual, empirical evidence that philosophers display the same (or similarly problematic) biases that non-expert philosophers do. To provide this evidence, one needs to adopt a direct strategy. Evidence provided by direct strategies is the focus of the next section.

[3] Rini (2014) provides an argument that cautions against drawing analogies between philosophy and some other disciplines. However, since the arguments here don't rely on analogies, these arguments can be safely ignored. That is, the arguments reviewed in this chapter state that because of the expertise (e.g., knowledge, skills) of philosophers, they will not fall prey to worrisome biases that we have documented. For example, because philosophers are experts, their personality should not influence their judgments.

Direct Strategies

Direct strategies are similar to indirect strategies in that they attempt to provide evidence that the kinds of expertise that philosophers are likely to have is not the right kind of expertise to eliminate the effects such as personality's relation to intuitions. But there is an important difference between direct and indirect strategies. Whereas indirect strategies try to draw connections between philosophy and other expertise domains, direct strategies try to show that philosophers have the same (or similarly problematic) biases, intuitions, or judgment tendencies as the philosophically naïve.[4] To do so, many researchers have started to document these biases, intuitions, or judgment tendencies in philosophers. There is gathering evidence that experts sometimes behave in much the same way as the folk. In this section, we review several attempts to directly assess the Expertise Defense. These studies provide evidence that philosophical training is likely to matter, at least in the sense that philosophical training appears to be related to different, more reflective cognitive styles, argument and evidence evaluation, and a tendency to gravitate to some philosophical positions. However, expertise does not generally remove or reduce the effects of at least some problematic biases in philosophers (perhaps similar to high cognitive abilities not being related to better decisions in chess). Consequently, to the extent that similar problematic biases are found in philosophical experts, the Expertise Defense fails.

To help contextualize direct strategies, it is important to have a sense of what kind of expertise philosophers are likely to have and if that is the relevant kind of expertise to support the expertise defense. To illustrate, take arguments about a specific kind of expertise—*moral expertise*. There have been a number of arguments about whether moral expertise can even exist, and if it does, what it is (pace our discussion of philosophical expertise). Some extreme views suggest that there is no moral expertise of any kind (Ayer, 1954; Broad, 1952; Ryle, 1957). Subsequent criticisms and defenses of moral expertise have been more nuanced largely by specifying the kinds of expertise that moral experts could have and the ways in which those moral skills could be actualized.

[4] Even if philosophers do not display the *same* biases, there could still be problematic associations with the contents of their intuitions and irrelevant factors. For example, there could be just as problematic reversals or slightly reduced effects for philosophers' intuitions (see, e.g., Weinberg, Alexander, Gonnerman, & Reuter (2012a)).

Peter Singer (1972) has detailed several ways in which one could become a moral expert (and these sentiments have been echoed by others, e.g., Archard (2011); Crosthwaite (1995); Hare (1989)). According to Singer, moral experts have special resources with respect to the following (similar to points 1–3 in the previous section):

1. Logical reasoning
2. Understanding of ethical theory and meaning of moral terms
3. Informed of relevant factual information
4. Time to think and reflect

Given these resources, it is often speculated that they will lead to more correct normative judgments (e.g., Singer writes "it would be surprising if moral philosophers were not, in general, better suited to arrive at the right, or soundly based, moral conclusions than non-philosophers" (1972, p. 117)). We see no reason to dispute that philosophers have access to any of the special resources indicated in 1–4. Indeed, most critics of moral expertise grant that philosophers have access to those resources.

What is critical for our understanding of the expertise defense is whether moral expertise goes beyond "descriptive" expertise to substantive expertise rather than being moral cartographers (Archard, 2011; Crosthwaite, 1995; Hare, 1989). "Cartographers" can explain the details of theories and meanings of terms, but may not be able to give better answers to substantive questions. To continue the analogy, cartographers might be able to tell one the best path and waypoints once a destination is chosen (e.g., Rome), but a cartographer cannot tell a person where they should go (e.g., Rome v. Venice). What is required for an adequate defense of the Expertise Defense is that moral experts typically come to true and correct conclusions—something Driver has called an Expert Judger. Expert judgers come to correct normative conclusions rather than simply drawing attention to morally relevant features and facts (e.g., determine whether moral objectivism is true rather than being able to articulate the theory and key terms). When there are disagreements among expert judgers, on Driver's view, one consults a meta-field expert that can reliably (even if not perfectly) determine which of two contradictory moral judgments is actually true. The critical question, then, for the Expertise Defense is whether there are any relevant meta-field experts to help alleviate the worries about personality's relation to philosophical judgments. We think there is good empirical reason to think that there is not, at least for many areas of philosophy. We turn now to empirical evidence to help support our claim that there are no meta-field experts in many areas of philosophy.

Philosophical Training and Cognitive Skills and Style

Recall one of the main points about expertise in general: Perhaps expert philosophers have some cognitive skills that non-expert philosophers do not. One way to show that philosophical training and expertise could insulate philosophical experts from the effects of extraneous factors is by showing that philosophical training increases some kinds of cognitive abilities likely to be related to philosophical thinking (hence, philosophical experts are always made and not born—the first important point about expertise in general). For example, perhaps philosophers have learned some metacognitive skills that make them less likely to engage in System 1 type errors. That is, philosophers may have some initial reactions to scenarios or thought examples, but that initial reaction may be corrected or attenuated because of their greater cognitive reflectivity. The correction may reduce biases associated with the folk because philosophical experts may have better logical reasoning skills, including metacognitive heuristics. If philosophical training gives philosophers some new abilities, then there might be good reason to think that the Expertise Defense may be successful. (We will save discussion of the third general point about expertise concerning experts having more nuanced knowledge structures than novices for later in this chapter.)

Some suggestive research indicates that philosophers have the ability to evaluate arguments and evidence in ways that people with other kinds of training may not. In one study, Kuhn (1991) performed a series of case studies on graduate students in philosophy. She found that those with graduate training in philosophy were generally better at evaluating arguments and evidence compared to experts in other domains (e.g., parole officers and teachers). While the sample of philosophy graduate students was small ($N = 5$), these results suggest that philosophers may have some skills that make their intuitions more robust against unwanted biases. These trained philosophers may be able to evaluate arguments and evidence in ways that allow problematic biases to be reduced or eliminated.

A different line of research suggests that perhaps philosophers have a unique, different, more reflective cognitive style than those without philosophical training. This reflective cognitive style may reduce the influence of extraneous factors. Cognitive reflectivity is a general way of thinking that is typified by careful, deliberate reasoning that can overcome intuitively compelling, yet wrong, responses (e.g., the Cognitive Reflection Task discussed above). Livengood, Sytsma, Feltz, Scheines, and Machery (2010) (see also Cokely and Feltz (2009a)) presented some evidence that those who have philosophical training across several levels of education had a more reflective cognitive style (see Fig. 5.1). These results suggest that philosophers have greater cognitive reflectivity than others, in particular non-philosophers or those who have had less exposure to philosophy. Hence, the differences in CRT for philosophers support the idea that philosophers may have abilities or skills that could reduce the influence of extraneous factors on their philosophically relevant intuitions.

The data concerning CRT and philosophical training is correlational so determining the causal direction is difficult. First, it could be that philosophical training causes increased cognitive reflectivity. Second, those who are higher in cognitive reflectivity may gravitate toward philosophy. Or, cognitive reflectivity and philosophical training may be caused by some other third variable. While Livengood et al. (2010) attempted to model whether the effect of philosophical training was causal, statistical tests were equivocal about the direction of causation, so given the currently available evidence there is no empirical way to prefer one direction of causation over the others. However, the direction of causation isn't required to help support the Expertise Defense. After all, the Expertise Defense simply states that philosophers, through their special training, skills, or

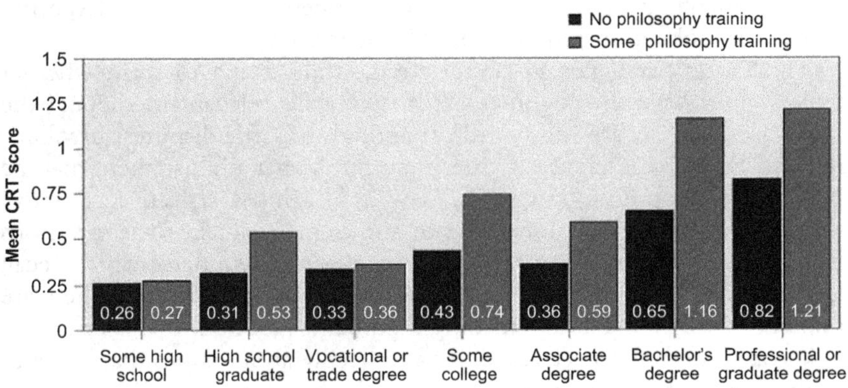

Fig. 5.1 Mean CRT scores across different levels of education for philosophers and non-philosophers (Livengood et al., 2010)

cognitive abilities, are not as prone to be influenced by extraneous factors. In this case, philosophers may have some special skills, perhaps antecedent to philosophical training, that make them less likely to be influenced by extraneous factors. It does not necessarily matter to the Expertise Defense the *way* that expert philosophers come to have the relevant expertise. Rather, all that is required is that expert philosophers in fact have the relevant expertise. To illustrate, consider an analogy. It does not matter if professional soccer players are naturally gifted or if they obtain their abilities through extensive training (which, of course, they do). All that matters to be a world-class soccer player is that one in fact has the relevant abilities regardless of how those abilities are obtained. So, philosophers having higher cognitive reflectivity may allow them to be reflective enough about the nuances of philosophical issues, thought examples, and intuitions and that may be enough to insulate the influence of extraneous factors on their intuitions.

Free Will, Extraversion, and Expertise

So far, the direct evidence for or against the Expertise Defense has only been suggestive. However, some of the evidence from direct strategies reviewed so far suggests that philosophers have cognitive skills that are different from those who are non-philosophers. Some of these skills or abilities may appear to support the Expertise Defense because philosophers appear to be more cognitively reflective and more skilled at evaluating arguments compared to non-philosophers. So far, the Expertise Defense is looking increasingly more likely to succeed.

The Direct Strategies so far reviewed, while trying to show whether philosophers have the cognitive abilities or skills relevant to support the Expertise Defense, are still not direct enough to settle disputes surrounding the Expertise Defense. Critics may think that philosophers may be better at evaluating arguments or may have a more reflective cognitive style, but those factors alone are not sufficient to shield their intuitions from the problematic aspects of irrelevant factors like personality. Recall the discussion of moral expertise. No one disputes that philosophers are good at constructing and defending arguments and nobody disputes that philosophers often reflect very long and deeply about issues that are core to their research program. Rather, the critic thinks that the key to the dispute is whether philosophers have systematically different contents of intuitions that serve as evidence for premises of arguments and whether these systematic differences are influenced by irrelevant factors. For example, extraverts may tend to have more compatibilist friendly intuitions, and they may be more likely to use the content of those intuitions as evidence for the arguments than introverts. Using their cognitive reflectivity and argument evaluation abilities, philosophers could create subtle and technically correct arguments based on the content of their intuitions. But, given the different content in their intuitions, extraverts and introverts could end up with quite different conclusions concerning the relation between determinism, free will, and moral responsibility. Of course, we suspect that these concerns aren't very compelling to defenders of the Expertise Defense. Apologists may simply reply that these skills or abilities do in fact give us reason to think that the Expertise Defense is correct. Is there a way that we can more efficiently settle the debate?

The answer is "yes." However, different methods are required. The currently reviewed methods only hint at the possibility that the Expertise Defense is correct. But, one could simply measure whether intuitions of philosophers are influenced in the same or similar ways to those of non-professionals. If the intuitions of philosophers are influenced by extraneous factors, then the Expertise Defense fails. If the intuitions of philosophers are not influenced by extraneous factors, then there is reason to think that the Expertise Defense may be correct.[5]

To start, there is some evidence that philosophers have some of the same biases that non-professionals have.[6] Schwitzgebel and Cushman (2012) conducted a series of experiments that attempted to show that an order effect in judgments that is common for some non-professional

[5] One might worry that the way that the differences between professional and lay intuitions is set up makes it impossible to determine whether the Expertise Defense is correct. This is because if there is no experimental evidence that philosophers' intuitions are not influenced, then it is a difficult inference to make that philosophers' intuitions are not so influenced. That is, the absence of evidence is not evidence for the absence of the effect. While this is a sensible worry, statistical tests can determine if expertise moderates effects (e.g., there is an effect for non-professionals but not for professionals). Moreover, if similar effects are found for experts, then that is evidence that the Expertise Defense fails.

[6] For other work along these lines, see Tobia, Buckwalter, & Stich (2013) and Horvath and Wiegmann (2022).

philosophers is also present in professional philosophers.[7] Theoretically, it should not matter to the truth of the content of an intuition whether scenarios are presented in one order versus another order. Order is an extraneous feature to the truth of the content of intuition. Consequently, it is widely agreed that the order effects constitute an irrelevant factor that could influence the content of the truth of intuitions.

In this case, Schwitzgebel and Cushman (2012) tested judgments about three different kinds of moral principles and three scenarios illustrating those principles. These scenarios and principles were drawn from the literature and concerned the Doctrine of Double Effect (e.g., diverting a trolley that kills one to save five versus pushing a large person that kills the large person but saves five), differences between actions and omissions (e.g., allowing to die versus killing), and finally cases of moral luck (e.g., a driver passing out and hitting a tree versus hitting and killing a person). They then gave these scenarios and principles to non-academics, non-philosopher academics, philosophers, and finally ethicists. Broadly, across all of the principles and scenarios, there were order effects for each of the three principles and for each of the three scenarios. In particular, the magnitude of the effect of order was roughly similar across all groups regardless of philosophical training.[8]

The results form Schwitzgebel and Cushman's studies are bad news for the Expertise Defense. Recall that one of the central empirical planks of the Expertise Defense is that philosophers, through their specialized training, skills, or cognitive styles, are less likely to display some of the effects of extraneous factors that non-philosophers display. However, Schwitzgebel and Cushman (2012) present data that this simply is not the

[7] There is some debate about whether the effects explored by Schwitzgebel and Cushman are genuine order effects. Rather, they may be more accurately characterized as updating effects. Order effects occur when participants receive all the same materials, are asked questions about those materials, and the different order of the materials influences the responses to those questions. Updating effects occur when one is presented with some materials, are asked questions, and then are presented with some more materials and are asked questions. The updating effect occurs when one has different responses to questions in those situations. It is uncontroversial that order effects are irrelevant to the truth of the answers, whereas updating effects could be. See Horne and Livengood (2017) for a more detailed discussion.

[8] If one thinks that expertise also means that one would live one's life in accordance with that expert knowledge (e.g., doctors more likely to live healthier lives), then one might expect ethical experts to lead better lives. However, across some paradigmatic tasks (e.g., voting, retuning library books) ethicists were no better than the average person (Schwitzgebel & Rust, 2009, 2014).

case—philosophers, even those who specialize in ethics—tend to display the same kinds of order effects to a similar degree as those who are non-professional ethicists. Hence, expertise does not typically appear to make the appropriate difference in the effect of an extraneous factor on intuitions as the Expertise Defense predicts.

While all of the studies that directly test the Expertise Defense are important and illuminating, they all suffer from a common shortcoming—they do not guarantee that the participants in the experiments are the relevant experts.[9] Often, expertise is identified simply by looking at a person's credentials or university degrees. However, those ways of identifying true expertise (as it was discussed above) is not always reliable (Ericsson & Lehman, 1996; Ericsson et al., 2007). For example, somebody who is a trained professional stockbroker may not be able to pick winning stocks at a greater frequency than non-stockbrokers. Many of the studies reviewed above used a credential-based approach (e.g., working in a philosophy department; self-identification of being a philosopher). Rather than relying on these kinds of credential-based approaches, a strong test of the Expertise Defense would be to test those who can demonstrate some elements of expertise. One of these necessary elements that we explored was the possession of superior objective knowledge about a field by asking people a set of questions to measure that objective knowledge.[10]

[9] Here, we are trying to give the defenders of the Expertise Defense the strongest position that we can think of. We do not think that critics of the Expertise Defense *require* that the relevant expertise is established in order to effectively call into question the Expertise Defense. Much depends on the goals of the critic. Philosophical expertise is often identified by credentials and restricting the range of that expertise may already be what a critic desires. That is, expertise may be restricted to a very, very narrow swath of philosophy (e.g., not just ethics in general, but rather to a small segment of the debate about moral objectivism). The restriction to verifiable experts to save the Expertise Defense may be "cleaving off the recliner to save the seat" as one of our manuscript reviewers helpfully suggested.

[10] One may object that knowledge of a domain is not sufficient for expertise in that domain. After all, one may know a lot about soccer without being an expert soccer player. We agree, but it does appear that for philosophical expertise, knowledge is a *necessary* condition. So, if one does not have the knowledge, one does not have the expertise. Relatedly, it is unknown if measuring one's knowledge is sufficient to measure one's expertise, but we believe that the two will be largely co-linear (i.e., if we measure knowledge we will also measure expertise). Training in philosophy, to the extent that it creates expertise, will be created by deep and prolonged reflection on the canonical works in the relevant field (i.e., the creation of durable, nuanced knowledge and representations of the problems). So, knowledge of a domain will likely translate into expertise in that domain.

Using an objective measure of philosophical expertise would allow us to identify those who have expertise in a specific philosophically relevant area versus those who do not. As reviewed above, research suggests that in some domains using credentials is not necessarily the best way to identify expertise. Philosophical expertise seems to be one of those domains. One reason that just being identified as "a philosopher" may not give a sense for what that philosopher's area of expertise is. Philosophy is a highly diverse field with many different topic areas (e.g., philosophy of physics, ethics, logic). Because of this diversity, one may be an expert in one area (e.g., mereology) but not an expert in another area (e.g., moral objectivism). The specialization of philosophy is not unique and is found in other disciplines as well. For example, you wouldn't necessarily go to a brain surgeon for heart surgery since they have different areas of expertise. Along these lines, somebody may be a professional philosopher yet not be an expert about free will (e.g., they may be an expert about mereology). We think defenders of the expertise defense can leverage these observations and argue that just being a "philosopher" does not necessarily mean that one is an expert about the relevant question. For example, even if extraversion predicted compatibilist intuitions for "philosophers" that does not mean that compatibilist would predict compatibilist intuitions for expert philosophers in free will. There may be something about that specific kind of training and expertise for free will experts that makes that relation go away.

All of this means that we are making it more difficult to adequately address the Expertise Defense. The reason why it is more difficult is because the set of relevant philosophical experts won't be identified by credentials or self-reports and the set will be much smaller than the set of "philosophers" (see, e.g., Stich (1998)). Given this specification and refinement of the Expertise Defense, we accept that we need evidence for the truth of a version of the following principle to claim that the Expertise Defense fails:

> (EQ) Philosophers' intuitions about hypothetical cases vary equally with irrelevant factors as those of non-philosophers. (Horvarth, 2010, p. 464)[11]

[11] J. M. Weinberg et al. (2010) refer to a similar principle they call the "ampliative inference" and inference where "patterns disclosed concerning the ordinary subjects to the predicted occurrences of those patterns in professional philosophers" (p. 332). See also J. Weinberg and Crowley (2009). Importantly, and as we've noted before, it could be the case that relations that do not "vary equally" could also be problematic for the Expertise Defense. But for the sake of argument here, we are willing to grant the "vary equally" requirement. And, we are also willing to grant that philosophical experts are only experts in a small area of philosophy.

The review about expertise provided above offers some good reasons for thinking that perhaps philosophical expertise makes a difference in a way to think that EQ is false. In some domains, experts just have higher quality intuitions (e.g., airplane pilots; chess players) and are not prone to the same kinds of judgment biases as non-experts (Ericsson & Lehmann, 1996; Ericsson et al., 2007). For example, grandmaster chess players have qualitatively different intuitions about game positions and risks than chess novices. Professional soccer players understand and can better predict what to expect during soccer play as compared to soccer novices. The idea is that through training and practice, these people just understand the problem space better, meaning their intuitions are better informed and calibrated. This could be no less true in philosophy.

To our knowledge, only one direct test of the Expertise Defense attempts to estimate philosophical expertise. This study involves extraversion's relation to compatibilist judgments in verified philosophical experts. We will treat this as our paradigmatic example.

It is not news that free will experts disagree about the correct answer to the compatibility question. There is even some debate about whether compatibilist or incompatibilism is the dominant view among free will experts. For example, some experts hold that "in contemporary discussions of free will, incompatibilists self-identify as the underdog" (Nichols, 2007, p. 261). However, other experts such as Robert Kane think "Compatibilism has surely become the dominant view among philosophers today" (1996, p. 12) and Derk Pereboom notes, "the demographic profile of the free will debate reveals a majority of soft determinists, who claim that we possess the freedom for moral responsibility, that determinism is true, and these views are compatible" (1995, p. 21). However, others think that there is evidence the most philosophers are incompatibilists.[12]

Whatever the professional landscape in free will is, we might ask if extraversion predicts that variation in expert intuitions about the compatibility question. There is reason to think that extraversion does. Because philosophers are also humans with personality, personality could be related to or influence philosophers' intuitions about free will and moral responsibility. If education for philosophers does not provide the right kinds of

[12] For a discussion of this, see http://gfp.typepad.com/the_garden_of_forking_pat/2005/02/how_many_battal.html. See also Bourget and Chalmers (2014) indicating that most philosophers gravitate toward compatibilism.

environments to create the relevant expertise, then we should see extraversion predicting intuitions about the compatibility question even in verified experts.

Schulz, Cokely, and Feltz (2011) explored whether extraversion predicted expert intuitions about the compatibility question. To do so, they asked a group of German participants to complete the extraversion subscale from the NEO-PI-R (Costa & McCrae, 1992, German version: Ostendorf & Angleitner, 2004). After completing the NEO-PI-R, participants read a standard determinism scenario. Participants were asked to rate their level of agreement with the Free Will questions on a scale from 1 (absolutely disagree) to 7 (absolutely agree). Finally, participants were presented with the following *Free Will Skill Test.*

1. "Well-known counterexamples for the PAP are called the Frankfurt cases. (true)
2. Arthur Schopenhauer said that there is definitely free will. (false)
3. Two important opinions in the debate about free will and determinism are called compatibilism and incompatibilism. (true)
4. PAP stands for the principle of alternate personalities. (false)
5. One frequently used argument for the freedom of choice is the experiment from Benjamin Libet. (false)
6. One well-known believer in free will was Jean Paul Sartre. (true)
7. The classical Trolley Problem is about two trains on a collision course. (false)
8. William James suggested that there could be soft determinism. (true)
9. One argument in the field of moral philosophy is Moore's open statement argument. (false)
10. The Stockholm's interpretation sees quantum physics as an argument against determinism. (false)"

After each question, participants had to indicate if they thought that the statement was true, false, or they did not know. A correct answer counted as one point, the wrong answer as a minus point, and "I don't know" as zero points, providing a correction for any participant guessing. The rationale was that if somebody was randomly guessing the answers, their total score would on average be 0.

There is good reason to think the Free Will Skill Test meets or exceeds the standards of classical test theory used for the development of many psychological and educational assessments (e.g., diagnosis, personnel selection). A validation study was conducted with 44 philosophy graduate students (age: *mean = 23, SD = 2.76; 16 females*). Among the moderately skilled sample in the validation study (*range: 0–9, mean = 2.9, SD = 2.2*), no distributional skew was observed. Further analysis indicated that the instrument had a very high test-retest correlation, $r = .99$ over short time intervals (10 to 30 minutes). The test had a Cronbach's Alpha of .75, which is above the conventional adequacy threshold for psychological instruments, providing further evidence of reliability and indicating high internal consistency. There was also evidence of convergent validity as the scores showed substantial correlations with free will relevant self-rated knowledge ($r = .5$), estimated number of papers read (*rho* = .49), lectures attended (*rho* = .33), and years one had been interested in the debate (*rho* = .39).

As predicted, extraversion—or more specifically warmth, which is an essential facet thereof—was systematically related to compatibilist intuitions. Importantly, with respect to the effect of personality on judgments of free will and moral responsibility, there was no reliable difference between folk and expert intuitions. The Free Will Skill Test predicted the compatibilist composite score ($M = 1.5$, *range* 0–10, $SD = 1.3$) explaining 9% of the variance (see model 1 in Table 2.1). Greater philosophical knowledge was associated with stronger incompatibilist intuitions. This suggests that as one is educated and learns more about the free will debate, one is likely to become more incompatibilist. However, warmth explained additional variance controlling for the Free Will Skill test score in a stepwise regression (Table 5.1). The full model explained 14% of the variance in participants' judgments (see model 4 in Table 2.1). Critically, when controlling for expert knowledge, warmth continued to predict a

Table 5.1 Explained variance of the different predictors. Please note that absolute r and F are indicated by Δr^2 and ΔF for models that are not stepwise (1, 2, and 4)

No.	*Model*	*Predictor*	*t(int)*	*p(t)*	Δr^2	ΔF	*p(F)*
1	Simple regression	Free will score	31	0.001	0.09	12.4	0.001
2	Simple regression	Warmth	3	0.002	0.06	8.3	0.005
3	Stepwise regression	(a) Free will score	31	0.001	0.09	12.4	0.001
		(b) Warmth	3	0.001	0.05	6.6	0.011
4	Full model	Both variables	3	0.001	0.14	9.8	0.001

moderate amount of unique judgment variance (about 5%; see Table 2.1), an estimated bias that is roughly equivalent in size to that observed among the folk.[13] The meta-analytic r^2 estimate from Chap. 2 was about .04, suggesting that we can have some confidence that expert's personality had similar relations to free will judgments as novices even given the small sample size.

These findings support some recent hypotheses suggesting that many people who are knowledgeable about the free will debate are incompatibilists. Results further indicated that both extraversion and expert knowledge were reliable, non-redundant, and not interacting predictors of judgment bias in a paradigmatic free will case. That is, extraversion biased expert and non-expert intuitions about the Compatibility Question to the same degree, suggesting that expertise does not eliminate or reduce the general compatibilism bias observed among experts. Training does change free will intuitions, but does not likely eliminate the effect of personality even for a group likely to be more cognitive reflective and knowledgeable. In short, free will training changes intuitions, but not always in the relevant ways. Consequently, at least for this paradigmatic example, the Expertise Defense fails.

The Difficult Death of the Expertise Defense

At this point, there is substantial reason to suspect that the Expertise Defense fails. Apologists no doubt have an arsenal of replies suggesting that it is *possible* that the Expertise Defense works. For example, Rini (2015) argues that we have some independent reason to think that the extraneous factors should be reduced or eliminated for philosophers because they are experts. Rini gives three possible explanations for problematic findings like the ones we have reviewed, all of which she argues are not threats to philosophical expertise. First, the philosophers who are polled are not really experts. Second, the philosophers who are polled don't really pay attention. Third, the philosophers who are polled haven't formulated responses to these kinds of scenarios because they are experts. That is, philosophers may have intentionally not formed responses to these scenarios because of various theoretical commitments. Finally, there could be some diachronic instability (e.g., with framing effects), but

[13] There was no interaction with expertise, gender, or age ($p > .3$).

philosophers could become aware of those effects and then use that knowledge to discount the justificatory role those intuitions play.

We think we've already made the case that given the evidence it is more plausible that philosophers do not have the right kind of expertise rather than any of the explanations offered by Rini. First, we can dismiss the "not experts" explanation. There is direct evidence that verifiable experts tend to display similar biases associated with personality in some paradigmatic instances (e.g., in free will). Second, experts likely *do* pay attention to the scenarios in surveys because they are about the very things that they have devoted a large amount of their lives to studying. In other words, they are likely intrinsically motivated to respond well. Similarly, if one is an expert in a domain, one most certainly has views about basic notions in the debate like determinism's relation to freedom and moral responsibility. Regardless, they seem to be sufficiently motivated to pass a basic knowledge test. Finally, at least with respect to personality, the problem isn't diachronic instability. The intuitions are diverse yet stable among different groups of individuals. As such, the diachronic explanation completely leaves the personality (and similar) findings untouched.

All of this does not call into question intuitions' use in all projects, and perhaps expertise has a valuable role to play in those other domains for the reasons already discussed. For example, perhaps one is interested in using intuitions in conceptual analysis (see Chap. 6 for a fuller discussion). Given one's access to intuitions along with greater cognitive reflectivity (or some other cognitive skill) and one's knowledge of the domain (or some other relevant knowledge) one may be able to skillfully construct a conceptual analysis that uses those intuitions as evidence. After all, we are not arguing that one's own intuitions are not indicative of one's own concept (on average), nor do we need to argue that philosophers do not have some genuine skills (e.g., skills of argument). Consequently, our arguments and data do not call into question (although they don't necessary support) those kinds of uses of intuitions or the Expertise Defense's role in supporting those kinds of philosophical practices. There may be other costs associated with taking this strategy (see Chap. 6) for other uses of intuitions. Nevertheless, those costs and benefits should be evaluated separately from the general evaluation of the Expertise Defense.

We think this puts critics of the Expertise Defense in a rhetorically strong position. If either indirect or direct strategies show that philosophers' intuitions vary as a function of extraneous factors in a similar way to folk intuitions, then the expertise defense is in trouble. If an indirect strategy is correct, then philosophers don't possess the relevant expertise to deflect the worrisome implications of the extraneous factors we have identified. If a direct strategy is successful, then philosophers' intuitions display similar biases as expressed by the folk or at least that intuitions show some systematic variation with personality. Given the amassing evidence that either an indirect or a direct strategy has merit in connection with the relative lack of evidence for the Expertise Defense, we can conclude that the Expertise Defense fails.

CHAPTER 6

The Philosophical Personality Argument

The core of this book revolves around establishing the soundness of the following argument:

The Philosophical Personality Argument (PPA)

1. "Philosophically relevant intuitions are used as some evidence for the truth of some philosophical claims.
2. Some differences in philosophically relevant intuitions used as evidence for the truth of some philosophical claims are systematically related to some differences in personality.
3. If philosophically relevant intuitions are used as some evidence for the truth of some philosophical claims and those intuitions are systematically related to some differences in personality, then one's endorsement of some philosophical claims is at least partially a function of one's personality.
4. Therefore, one's endorsement of some philosophical claims is at least partially a function of one's personality." (A. Feltz & Cokely, 2012a)

This chapter attempts to establish the truth of premises 1 and 3. Chapters 2–4 were devoted to establishing the truth of premise 2. For those who are more interested in the empirical results concerning philosophical intuitions, refer to those chapters. For those who are only

A. Feltz, E. T. Cokely, *Diversity and Disagreement*,
https://doi.org/10.1007/978-3-031-61935-9_6

interested in the more sustained argument about what practical implications the PPA has, please skip to Chap. 7. In this chapter, we argue that the PPA has important implications for some philosophical practices. In particular, our goal is to not only connect the findings from earlier chapters to provide evidence for 1 and 3, we will go on to argue that the conclusion represented in statement 4 means that we should not use (or at least significantly discount) intuitions in some major philosophical projects. This view has been characterized as a "restrictionist" view (Alexander & Weinberg, 2007). Along the way, we will consider and answer some objections to the PPA.

Philosophical Projects

To be clear from the beginning, we do not think the results or our arguments from previous chapters have restrictionist implications for all philosophical projects. Our restrictionist target is much more, well, restricted. We do not think that the empirical results that we have reported in Chaps. 2–4 call into question all of philosophy or all philosophical projects. Rather, we think that our results support a limited form of restrictionism only for some philosophical projects. To illustrate, many have noted and discussed the implications of empirical data for a particular project in philosophy—conceptual analysis (see below for a fuller discussion of conceptual analysis) (Kauppinen, 2007; Ludwig, 2007, 2010). But that is not the only general philosophical project one could engage in. As Sosa notes,

> [T]he use of intuitions in analytic philosophy, and in philosophy more generally, should not be tied to conceptual analysis. Consider some of the main subjects of prominent debates in analytic philosophy: utilitarian versus deontological theories in ethics, for example, or Rawls's theory of justice in social and political philosophy, or the externalism/internalism debates in epistemology, and many others could be cited to the same effect. These are not controversies about the conceptual analysis of some concept. They seem moreover to be disputes about something more objective than just our individual or shared concepts of the relevant phenomena. Yet they have been properly conducted in terms of hypothetical examples, and intuitions about these examples. The objective questions involved are about rightness, or justice, or epistemic justification. ((Sosa, 2007b, p. 59), see also (Sosa, 2007a, 2009))

We agree with the general idea that Sosa presents. Conceptual analysis is not the only general philosophical project philosophers engage in. For example, other projects could be normative projects that may be primarily concerned with the identification of how one ought to behave or what one ought to believe.

Along these lines, Stich and Tobia (2016) argue that empirical results concerning intuitions have different implications for different philosophical projects. They point out three types of projects in contemporary philosophy (that do not exhaust the projects in contemporary philosophy): (1) projects in conceptual analysis, (2) Neo-Platonic projects, and (3) normative projects. Conceptual analysis aims to provide an analysis of philosophically relevant concepts. For example, the Justified True Belief (JTB) account of knowledge could be understood as an analysis of the concept of "knowledge" (see for some related examples Audi (2011)). On this account, one knows some proposition *p*, if and only if (1) *p* is true, (2) one believes that *p*, and (3) one has good evidence for *p*. Neo-Platonic projects attempt to discover the truth about some non-conceptual or non-linguistic philosophically relevant phenomenon. For example, one attempts to discover what knowledge *is* or what intentional action *is* and not what some people's *concepts* of knowledge or intentional action are. To put the distinction in other (somewhat misleading) terms, conceptual analysis deals with things in the head and Neo-Platonic projects deal with things in the world and universe (Stich and Tobia (2016) refer to this as the distinction between *mental* and *extra-mental* projects). Finally, normative projects address how we ought to be. Two of the most prominent fields in this project are epistemology and ethics. Epistemology largely deals with what we ought to believe and ethics largely deals with how we ought to act.

Few restrictionists want to call into question *all philosophical uses of intuitions*. Rather, restrictionists are often worried about certain classes of intuitions used in the service of certain projects (J. Weinberg, 2006). In this chapter, we will be mostly concerned with the impact the PPA has for Neo-Platonic (i.e., extra-mental) projects. In short, if the PPA is sound, then many Neo-Platonic projects run the risk of not being able to be reliably conducted given the current, dominate approaches to Neo-Platonic projects. But first, we begin with a sketch of a defense of each of the premises in the PPA.

The Truth of the PPA's Premises

Premise 1

The PPA is valid but are all the premises true? We think they are. Premise 1 describes what has been called the "practice of philosophy" (Alexander & Weinberg, 2007). As Hilary Kornblith puts it, "Most philosophers do it openly and unapologetically, and the rest arguably do it too, although some of them would deny it. What they all do is appeal to intuition in constructing, shaping, and refining their philosophical views" (1998, p. 129). That is, intuitions are often used as evidence for some philosophical claim (Bealer, 1998; Pust, 2000, 2001). For example, "intuitions are supposed to function like observations" in empirical sciences (Sosa, 2007a, p. 106) (see also Sosa (2009)).

As these quotes indicate, it is commonly thought by many contemporary philosophers that intuitions are crucial pieces of evidence for some philosophical claims. However, unpacking exactly what it means for intuitions to be used as evidence for some philosophical claims is tricky. On the face of it, one might think that there are at least three issues that need to be addressed in order to have confidence that Premise 1 is true: (a) one must understand what an intuition is, (b) one must understand how those intuitions are used as evidence, and (c) it must be true as an empirical claim that intuitions are used as evidence. Currently, debates exist about each of (a)–(c). Here, we attempt to address some of these issues.

What is an intuition? Perusing the philosophical literature on intuition does not help much. Consistent with the general theme of this book, there is a plurality of notions of what intuitions are (for a fuller discussion of how intuitions have been characterized, see Feltz and Bishop (2010)). According to David Lewis, "'intuitions' are simply opinions" where "some are commonsensical, some are sophisticated; some are particular, some are general, some are firmly held, some less. But they are all opinions" (Lewis, 1983, p. x). According to Peter van Inwagen, for example, "'intuitions' are simply beliefs—or perhaps, in some cases, the tendencies that make certain beliefs attractive to us, that 'move' us in the direction of accepting certain propositions without taking us all the way to acceptance" (Van Inwagen, 1997, p. 309). According to Ernest Sosa, an intuition is "a representationally contentful conscious state that can serve as a justifying basis for belief while distinct from belief, not derived from certain sources, and possibly false" (2007b, p. 57). George Bealer argues that intuition is a "sui generis, irreducible, natural (i.e., non-Cambridge-like) propositional attitude that occurs episodically" and are distinct from "physical intuitions, thought experiments, beliefs, guesses, hunches, judgments, common sense, and memory…not reducible to inclinations, raisings-to-consciousness of non-conscious background beliefs, linguistic mastery, reports of consistency; and so forth" (1998, p. 213).

Many philosophers who write about intuitions claim that they are non-inferential. For example, Lisa Osbeck claims that "the salient feature common to various accounts of intuition is its non-inferential status" (2001, p. 119). Alvin Goldman and Joel Pust "assume, at a minimum, that intuitions are some sort of spontaneous mental judgments. Each intuition, then, is a judgment 'that p', for some suitable class of propositions p" (1998, p. 179). But some naturalistically inclined philosophers take intuitions to be the result of some inferential process. Michael Devitt claims that "intuitive judgments are empirical theory-laden central-processor responses to phenomena, differing from many other such responses only in being fairly immediate and unreflective, based on little if any conscious reasoning" (2006, p. 491). Hilary Kornblith argues that intuitions "are corrigible and theory-mediated" (2002, p. 13).

Philosophers disagree about whether intuitions are commonsense, untutored judgments or whether they can arise (non-inferentially) after considerable learning and reflection. So L.J. Cohen contends that "an intuition that p is...just an immediate and untutored inclination, without evidence or inference, to judge that p" (1981, p. 318). On the other hand, Laurence BonJour takes intuitions to be "judgments and convictions that, though considered and reflective, are not arrived at via an explicit discursive process" (1998, p. 102).

Some philosophers believe that intuitions come with a characteristic feeling or conviction that what is intuited is true. Guy Claxton thinks that intuition "comes to mind with a certain aura (or even conviction) of 'rightness'" (1998, p. 217). Stephen Hales thinks that "to have an intuition that A is for it to seem necessarily true that A" (2000, p. 137). But as we have already seen, those philosophers who take intuitions to be beliefs do not suppose that intuitions *must* come with these sorts of seemings, although they can include an inclination to accept a belief.

There is a cacophony of views about intuitions. Feltz and Bishop (2010) suggest that we can capture some of the variation in views about intuitions in terms of the following menu:

Menu A: Choose one each from the As, the Bs, and the Cs.

A1. Intuitions are beliefs or inclinations to believe.
A2. Intuitions are sui generis propositional attitudes.
B1. Intuitions are inferential judgments.
B2. Intuitions are non-inferential judgments.
C1. Intuitions include only untutored judgments.
C2. Intuitions include tutored and untutored judgments.

This menu defines eight different views about intuitions (in terms of the various possible combinations of As, Bs, and Cs).

So, what do we mean by intuition? The answer may be somewhat anti-climatic. We take no substantive position on which combinations of A-C intuitions are because we don't take it as that important what the psychological states of the intuitions are—e.g., if intuitions are different from judgments, if intuitions can be inferential or not, etc. This is a debate we think we can safely avoid because the PPA does not need to assume anything substantive about the nature of intuitions. As we hope will become clear, what we take as important about intuitions is their contents and not their psychological properties. For example, when one has a Gettier intuition, it is not the fact that one has an intuition that is important. Rather, what is important is that one has the intuition that has *as its content* that the person does not know.

Another crucial notion is what it means for intuitions to be *used* as *evidence*. There are several philosophical conceptions of evidence (Achinstein, 2000). For our purposes, not much hangs on the correct philosophical account of evidence. Premise 1 is consistent with many philosophical conceptions of evidence. To illustrate, take the following popular analysis of evidence: X is evidence for (against) Y if and only if X makes the truth of Y more (less) probable (Achinstein, 1994; Maher, 1996). This analysis of evidence can be easily modified so that intuitions are evidence for some philosophical claims: an intuition (or cluster of intuitions) *I* is evidence for (against) some philosophical claim *C* if and only if *I* increases (decreases) the probability that *C* is true.[1] For example, intuitions about Gettier cases typically are thought to decrease the probability (perhaps to 0) that the JTB account of knowledge is true. Hence, intuitions about Gettier cases are evidence against the JTB account.

However, having an *I* for or against *C* is not sufficient for *I* to be *used* as evidence. It is widely accepted that intuitions are defeasible evidence for the truth of some philosophical claims (Bealer, 1998; Sosa, 2007a, 2007b, 2009). Just as one could have visual intuitions that the moon is larger than

[1] Goldman and Pust argue something similar, "Mental states of type *M* constitute a basic evidential source only if *M* states are reliable indicators of the truth of their contents (or the truth of closely related contents), at least when the *M* states occur in *M* favorable circumstances" (1998, p. 180).

the sun without *using* that intuition in one's astronomy, one could have a philosophically relevant intuition without *using* that intuition as evidence in one's philosophical theory. One could discount intuitions in a number of ways. To illustrate, one favored method of generating philosophical theories is wide reflective equilibrium (Bealer, 1998; Daniels, 1979; Goodman, 1955). Wide reflective equilibrium counsels revising a theory if it conflicts with a deeply held intuition shared by many (along with appropriate background theories). By a process of mutual adjustment between intuitions and background theories, one settles on a theory. In creating equilibrium between intuitions and principles, some intuitions could be discarded. For example, the intuitions about the moon could be discarded given background theories and other evidence. Those discarded intuitions do not increase or decrease the probability that some philosophical claim is true. In such situations, those discarded intuitions are no longer *used as evidence* for the truth of a philosophical claim. As a result, we understand *I used as evidence* when *I* enters into a justificatory process where *I* figures into the probability that a philosophical claim is true.

Of course, whether reflective equilibrium is the right method for philosophical theorizing (Stich, 1998) or if intuitions should play an evidential role is not uncontroversial (Cappelen, 2012; Deutsch, 2010, 2015; Timothy Williamson, 2007). The actual correctness of either is not our main concern here. What is important is that this account captures how many philosophers in fact treat the contents of intuitions. A perusal of the philosophical literature reveals the contents of intuitions used in ways consistent with the above analysis of evidence where the contents of intuitions are used as evidence to decrease or increase the probability of some philosophical claim. We find Chinese rooms (Searle, 1980), Swamp Men (Davidson, 1987), counterfactual interveners (Frankfurt, 1969), and strangely wired video games (Bratman, 1984) that are meant to generate an intuition in the reader. The content of these intuitions is then used as evidence either for or against philosophical claims. Many philosophers take the contents of these types of intuitions to be valuable or even irreplaceable parts of philosophical practice (Bealer, 1998; Daniels, 1979; Jackson, 1998; Ludwig, 2007; Pust, 2000, 2001; Sosa, 2007b). It is this practice that many empirically minded philosophers have been interested in (Alexander & Weinberg, 2007; E. T. Cokely & Feltz, 2009a, 2009b, 2011; J. S. Miller & Feltz, 2011; Stich, 1990, 1998; J. Weinberg, 2006; J. Weinberg et al., 2001). So, while not all philosophers take intuitions to be central evidence for philosophical claims, many philosophers think that intuitions provide some important evidence for many philosophical projects.

To summarize, it is largely irrelevant to our argument what intuitions are. However, we think that it is also clear that the contents of intuitions are often thought to be, and are often actually used as, evidence for some philosophical claims. If the contents of intuitions are used as evidence for or against some philosophical claims, then Premise 1 is true.

Premise 2

We have presented evidence in Chaps. 2–4 that personality predicts some philosophically relevant intuitions. For example, some intuitions about free will and moral responsibility are predicted by the heritable personality trait *extraversion* (see Chap. 2) (Andow & Cova, 2016; Cokely & Feltz, 2009a; Feltz, 2013; A. Feltz & Cokely, 2008, 2009; Feltz & Millan, 2015; Feltz, Perez, et al., 2012b; Nadelhoffer et al., 2009), as are some intuitions about intentional action (see Chap. 3) (Cokely & Feltz, 2009b; Feltz, Harris, et al., 2012a). Similarly, some intuitions about moral objectivism are predicted by the heritable personality trait *openness to experience* and some intuitions about virtue are predicted by the personality trait *emotional stability* (see Chap. 4) (Cokely & Feltz, 2011; Feltz & Cokely, 2008, 2012b, 2013b). This research indicates that personality traits can, at least in part, predict a variety of philosophically relevant intuitions in a variety of philosophically relevant domains.

At this point, one may wonder about the strength of the empirical evidence that personality predicts philosophical disagreement. While sustained defenses occurred in the previous three chapters, here we will briefly consider the recent "replication crises" that has pre-occupied researchers in experimental philosophy and psychology more generally. This preoccupation is justified in the light of recent findings of fraud and high-profile (if disputed) attempts at replications (Open Science, 2015). Indeed, even in experimental philosophy there have been results that have been difficult to replicate (Seyedsayamodst, 2015). One might think that personality's relation to philosophically relevant judgments could be the same—namely, that there is in fact no relation between personality and one's philosophical judgments. Indeed, some have argued that some of the paradigmatic examples of personality and philosophical judgments (e.g., between extraversion and free will) have not been replicated (Mortensen & Nagel, 2016). Thus, either given the failure of replication or the possibility of failure, we should have little confidence that Premise 2 is true.

Science involves risk, so premise 2 could turn out to be false, of course. However, the mere possibility that the relations aren't real is nothing all that new or threatening to Premise 2. Rather, the question is whether the on-balance evidence suggests that Premise 2 is true. Failure to replicate a few or even a substantial number of times does not in itself call into question the reliability of the relation between personality and philosophically relevant intuitions (see, e.g., the meta-analysis in Chap. 2). There are any number of explanations why experiments could fail to replicate—explanations too numerous to enumerate here. What is important is that estimating the reality of the underlying relations that is difficult to determine with one or two experiments (Cumming & Calin-Jageman, 2017). That said, we think the body of work speaks for the reality of the relations.[2] While being certain of Premise 2 is not to be had, there is growing evidence that the relations are in fact robust and reliable, persisting independent of wide differences in abilities, presentation formats, ages, cultures, education levels, expertise, and general demographics. The weight and consistency of the evidence suggests we'd be foolish to expect that the evidence will not continue to mount. In other words, it is theoretically possible there is no relationship between personality and philosophical intuitions, but at this point that possibility appears extremely unlikely.

Premise 3

Premise 3 also appears to be true. An intuition is used as evidence when it enters into one's justificatory process (e.g., wide reflective equilibrium), and as a result, the content of the intuition increases or decreases the probability that a philosophical claim is true. If the contents of intuitions are used as evidence, then the view that one ends up endorsing is a function of the contents of those intuitions. In addition, gathering evidence suggests that some philosophically relevant intuitions often used as evidence are systematically related to global personality traits. Intuitions in the domains we have documented (e.g., free will, intentional action, ethics)

[2] As more data gathers, meta-analytic techniques like we have used in previous chapters could be applied to help understand the strength and extent of the relations between personality and philosophically relevant intuitions. To date, there simply are not enough studies to meaningfully meta-analyze to determine the strength of possible moderating variables.

are influenced by personality and continue to be used as evidence for some philosophical claims. Hence, if a widely endorsed method of philosophical inquiry is used and philosophically relevant intuitions vary as a function of personality, then the philosophical view one ends up holding will be at least partially related to one's personality.

Implications of the PPA

We have tried to be careful in arguing that the implications of the PPA are dependent on what projects one is engaging in. In this section, we distinguish three types of projects one could engage in: Neo-Platonic projects, conceptual analysis, and normative projects (Stich & Tobia, 2016). The implications of the PPA for each of these projects is different, as we detail in what follows.

Neo-Platonic Projects

Neo-Platonic (or "extra-mental") projects attempt to discover the non-conceptual, non-linguistic truth of the relevant philosophical phenomenon by using rational reflection along with relevant intuitions. On this project, through rational discourse, we can come closer to achieving or approximating the truth. The PPA suggests that some agreement or disagreement in Neo-Platonic projects is not solely a function of *purely* rational arguments aimed at a progression toward the truth. Rather, some features irrelevant to the truth of the content of the intuition (e.g., personality traits) may be driving mechanisms of philosophical agreement or disagreement.

We think that there is wide agreement that personality is not related to truth for almost all Neo-Platonic projects. As we mention below, that would be like claiming that one's personality is related to the truth about how many electrons copper has or how many moons Jupiter has. For these reasons, the PPA is another argument for *restrictionism*. Restrictionism holds that "the results of experimental philosophy should figure into a radical restriction of the deployment of intuitions as evidence" (Alexander & Weinberg, 2007, p. 61) because "it involves deploying a source of putative evidence that is sensitive to non-truth-tracking factors" (J. M. Weinberg et al., 2010, p. 332). There are many results from psychology and related disciplines that suggest that many intuitions that could be or have been used in Neo-Platonic projects are associated with extraneous features.

These extraneous features thereby call into question whether those intuitions are reliable guides to the Neo-Platonic truth.[3] For example, some philosophically relevant intuitions in some paradigmatic cases vary with respect to socio-economic status (Haidt et al., 1993), culture (Huebner, Bruno, & Sarkissian, 2010; Machery et al., 2004; J. Weinberg et al., 2001), the presentation order of scenarios (Feltz & E.T. Cokely, 2011; Swain, Alexander, & Weinberg, 2008),[4] and one's perspective (Alexander, Betz, Gonnerman, & Waterman, 2018; A. Feltz, Harris, et al., 2012a; Nadelhoffer & Feltz, 2008). Several theorists contend that associations such as these call into question the truth of the content of those intuitions (Alexander & Weinberg, 2007; Horvarth, 2010; Sinnott-Armstrong, 2008; Stich & Tobia, 2016; J. Weinberg, 2006; J. Weinberg et al., 2001). The actual mind-independent truth about determinism's relation to free will and moral responsibility should not depend on the order in which questions about that relation are presented (just as, e.g., the weight of a 10 pound bar does not vary just because one was previously holding a feather, although people may tend to judge that weight differently after holding a feather). Personality is like these factors. If one has the intuition that free will and moral responsibility are compatible with determinism, it would at least be odd to defend the truth of the content of the intuition by appealing to the fact that one is extraverted. It would be like saying that one's personality is relevant to whether the atomic number of gold is 79.5. Yet some of those free will intuitions are predictably related to personality. As such, those intuitions may be related to extraneous factors. Given that there is currently no argument that successfully allows us to always prefer one set of intuitions to another set (introversion and extraversion seem to be equally irrelevant to the truth of whether determinism is compatible with free will and moral responsibility. See below for a fuller discussion of objections), one should not dismiss one set of intuitions. To the extent that these intuitions vary with irrelevant factors such as personality, it does not appear that some philosophical disagreements about some Neo-Platonic projects are solely a function of rational disagreement (see

[3] For a different approach about how to interpret "reliable" here, see Machery (2017).

[4] Horne and Livengood (2017) have a compelling argument that the results of these studies are not the result of genuine order effects but rather may reflect updating effects. Neither genuine order effects nor updating effects may be vicious. For example, the former depend on the magnitude and extent of the order effect on judgments and the later may actually be epistemically virtuous because one uses new information to update one's attitudes, beliefs, or judgments.

(Sommers, 2012)). The use of any source of evidence in Neo-Platonic projects that does not only track the truth should be restricted. Hence, the use of intuitions that vary with personality should be restricted in Neo-Platonic projects.

But some care needs to be taken in interpreting restrictionist implications of the PPA. To illustrate, take Horvarth's (2010) formulation of the restrictionist's Master Argument:

> "(A) Intuitions about hypothetical cases very with irrelevant factors.
> (B) If intuitions about hypothetical cases vary with irrelevant factors, then they are not epistemically trustworthy.[5]
> (C) Intuitions about hypothetical cases are not epistemically trustworthy." (p. 448)

It seems like the PPA nicely complements the Master Argument. But, one worry is that the conclusion of the PPA does not support (A). One might think that personality could be relevant to the truth of particular philosophical claims. To take just one example, Prinz (2007) argues that the truth of some moral claims is essentially related to one's emotions. For example, when one says "Eating your dead relatives is wrong," the truth of that claim essentially involves a sentiment (i.e., disposition to experience disapprobation) in the speaker. Without the sentiment toward eating a dead relative, what the speaker says is false on Prinz's account. On the assumption that personality can influence what sentiments one is likely to have, it seems that it is possible that personality *is* reliably related and relevant to the truth of some philosophical claims (e.g., "Eating your dead relatives is wrong").

We do not dispute that personality could be related to the truth of *some* philosophical claims (in fact, some of our main claims *require* it). However, our main point is about whether personality is related to the truth of the content of intuitions for Neo-Platonic projects. Here, it is illustrative to see two different claims that might be made on Prinz's view:

> (M) An action has the property of being morally wrong (right) just in case there is an observer who has a sentiment of disapprobation (approbation) toward it. (2007, p. 92)
> (N) Eating your dead relatives is wrong.

[5] A piece of evidence is trustworthy if one can detect and correct for errors in those pieces of evidence (J. Weinberg, 2006).

Presumably, personality can be importantly and reliably related to the truth of (N) but not the truth of (M). One plausible explanation is that (M) is an example of a *Neo-Platonic* claim whereas (N) is a *normative* claim. (M) is a "Metaphysical Thesis" about the nature of a philosophically relevant phenomenon indicating a requirement for something to be morally wrong or right (Prinz, 2007, p. 92). The truth of Neo-Platonic claims like (M) is not supposed to be related to one's personality. However, (N) is a normative claim. The truth of (N) could be importantly related to one's personality (see Chaps. 4 and 5). Somebody who is extremely low in emotional stability might think that (N) is true whereas people who are high in emotional stability may be more likely to think (N) is not true. Even if one argues that (N)-like statements support the generalization to (M), there could be another who does not think that the (N)-like statements support (M) as a mind-independent fact about the world. If people who are high and low in emotional stability have qualitatively different intuitions about (M), at least one of them is wrong. As a result, personality could be critically important for the truth of some *normative* claims, but it does not appear that personality is important or relevant for the truth of many (if not all) *Neo-Platonic* claims.

Defenders of traditionally conducted Neo-Platonic projects may ask "is the PPA problematic for Neo-Platonic projects that use intuitions as evidence?" The defender may argue that on the conception of evidence we have been using, having an *I* for (or against) a *C* is not sufficient for *using I as evidence for (or against) C*. Once philosophers learn that some of their intuitions are systematically related to some personality traits, they may appropriately discount those intuitions and not use them as evidence. And, the argument goes, excluding or discounting some intuitions is a natural part of philosophical practice. As a result, the PPA should be of no concern for those using intuitions as evidence because almost all philosophers incorporate relevant empirical evidence into their philosophizing already. Philosophers can admit that learning some intuitions are systematically related to personality is important information to incorporate into theorizing, yet that information does not call into question Neo-Platonic theorizing based on intuitions as the PPA suggests.

While such a position is possible, we think it would be a major concession to restrictionists for at least two reasons. In one sense, consulting closely with empirical psychology may not be odd or new for those working on some Neo-Platonic projects (Sosa, 2007a). Some have argued that the conditions under which one is free or morally responsible may be helpfully informed by empirical science (Nahmias, 2007). Empirical psychology may tell us under what conditions one lacks freedom-relevant control over a behavior (e.g., one's glucose is too high or low, cf. Baumeister (2008)). However, in another sense, consulting with the empirical sciences would be important and new for Neo-Platonic projects. In this sense, it is not merely that empirical evidence has *some* role to play in assessing some philosophical claims. Empirical evidence has a role to play in evaluating the *intuitions* deployed in Neo-Platonic projects. For example, is one's glucose being too high or too low relevant to the control required for free will and moral responsibility? The answer to this question may depend on what intuitions one has about the particulars of the case involving glucose, and those intuitions may be systematically influenced by personality. It isn't odd to think that glucose's effect on behavior is relevant. It is odd to think that one's personality influences one's thinking about the importance of glucose's effect on behavior.

We know of very few philosophers (if any) who discard intuitions because the intuitions are likely influenced by the type of personality they have. For example, compatibilists and incompatibilists neither reject their own intuitions nor the intuitions of others because of personality's relation to those intuitions. But the PPA suggests that these intuitions sometimes need to be dramatically discounted as evidence in some Neo-Platonic projects. More generally, the PPA suggests that in order to have confidence in intuitions used in Neo-Platonic projects, we need to have a deeper psychology of philosophical intuition (see Stich and Tobia (2016)) and philosophical expertise. Hence, the PPA would constitute a significant change to the way some philosophers go about doing Neo-Platonic projects.

Second, philosophers who think that intuitions related to personality can be discounted find themselves in a precarious philosophical position: The viability of their approach depends critically on empirical science. That is, at the end of the day, for their position to be viable there had better not be large and systematic relations of one's personality with the intuitions that one has. There is already evidence that personality is related to intuitions in problematic ways for Neo-Platonic projects. We suspect that the evidence indicating that philosophically relevant intuitions are systematically related to personality (and other stable individual differences) will continue to grow. If our suspicion turns out to be right, then these philosophers are hostage to empirical results. What if philosophically relevant intuitions in a substantial number of fields are systematically related to personality traits? If that is the case, then the defender of intuitions in Neo-Platonic projects would be committed to barring those intuitions as evidence for Neo-Platonic projects. But if a sufficient number of intuitions are thereby barred, there will not be a sufficient amount of evidence to theorize upon. These considerations lead to a rather striking suggestion: In their current form, many intuition-driven Neo-Platonic projects simply should not, and possibly cannot, be done.

It is important to note that the worry posed by the PPA is not merely a skeptical hypothesis. We aren't just positing the *possibility* that personality is systematically related to intuitions used in Neo-Platonic projects. Rather, the PPA (and the evidence reviewed in the previous chapters) indicates that this possibility is likely actual. What we currently know is sufficient to justify the worry that many Neo-Platonic projects based on intuitions are in trouble. What we need to do, then, is to investigate the extent to which intuitions in the relevant Neo-Platonic project are associated with personality traits (and other similar psychological variables). And that means that almost all philosophers engaged in Neo-Platonic projects using intuitions as evidence need to consult empirical evidence about their own (and others') dispositions and intuitions more closely (i.e., via scientific methods).

Given these considerations, we favor the following principle:

> (E) Empirical evidence should play a substantial role in many philosophical projects.

We take (E) to be the general attitude of experimental and empirically minded philosophers. The empirical evidence is both evidence relevant to the Neo-Platonic project and evidence about intuitions deployed in those projects. Many philosophers engaged in Neo-Platonic projects reject (E) in some way. One could reject (E) by simply ignoring (E). However, we think following principle is true:

> (R) Most fully adequate philosophical views should take into account all available, relevant evidence.[6]

[6]Other prominent philosophers hold a similar principle. For example, Kaplan (2000) writes "that an epistemological hypothesis must (like any other) be held accountable to all the evidence" (p. 301). We thank a reviewer for bringing this to our attention.

Indeed, we think that (R) is entailed by reflective equilibrium—one of the core philosophical techniques that use intuitions as evidence. Ceteris paribus, if a philosophical theory does not account for available and relevant evidence, then that theory will at least in some cases (and perhaps in most cases) be inferior to one that does take into account that evidence. Personality's influence on intuitions is available and relevant. Therefore, ignoring personality's influence on philosophically relevant intuitions is in most cases not tenable.

Conceptual Analysis

Generally, whether intuitions ought to be trusted depends on what intuitions are to be trusted for. Our main target in this chapter is some Neo-Platonic projects. However, we will briefly comment on two other projects in which intuitions have been used as evidence—conceptual analysis and normative projects (Stich & Tobia, 2016). Indeed, we want to stress that the arguments that we have provided do not have restrictionists implications for intuitions in all philosophical projects. Philosophy is a diverse field with many different projects, some of those projects can use intuitions and perhaps even should use them as evidence. One is in conceptual analysis. One prominent approach to conceptual analysis attempts to provide a set of individually necessary and jointly sufficient conditions for something to be classified under that concept. For example, an analysis of "knowledge" attempts to give the individually necessary conditions that if a person lacks, then the person does not know, and the jointly sufficient conditions specifying that if a person satisfies all those conditions, then one knows. As has already been mentioned, the Justified True Belief account of knowledge conceived of as an analysis of the concept provides the following individually necessary and jointly sufficient conditions for knowing: (1) a person believes that *P*, (2) *P* is true, and (3) the person has justification for believing that *P*.

If people have different intuitions about knowledge, does that mean conceptual analysis about knowledge cannot be done? Not necessarily. Our view is consistent with the view expressed by Machery (2017) (who also does not think that all intuitions ought to be restricted). He holds a view of "naturalized conceptual analysis." On this view, conceptual analysis could still be done but it should pay close attention to empirical results indicating that different people have different intuitions. There are many ways that using these empirical results about intuitions still could be used in a valuable evidential role. To illustrate one way, perhaps the diversity of intuitions simply means that different (groups of) people have different concepts. There is some evidence that at least for some philosophical concepts the multi-concept view is true. Recall some of the discussion about intentional action intuitions from Chap. 3. There, we reviewed some evidence that suggests that there could be two or three different concepts of intentional action that are responsible for the pattern of results observed in the side-effect effect (Cushman & Mele, 2008; Mele & Cushman, 2006; Nichols & Ulatowski, 2007). Additionally, our evidence suggests that we can at least partially predict who is likely to have those concepts with extraversion. It could be, then, that extraverts and introverts have different intentional action concepts. Maybe something similar is happening in at least some other philosophical domains. While we don't currently have specific evidence to support the following claim, it could be that some people have different concepts (and corresponding intuitions) in all of the areas we have discussed (free will, some areas of ethics, intentional action) (Sosa, 2007a, 2009). If all of this is correct, then conceptual analysis can still be done with the important caveat (or restriction) that the conceptual analysis is not about the *only* concept (e.g., *the* concept of intentional action).[7]

We do not take a stand whether the multiple concepts approach is tenable. There do seem to be at least some potential costs if one adopts a view that there are potentially many concepts in a specific philosophical area. Some may find it objectionable and costly to concede that the analysis they are providing is not about *the only* concept in that philosophical domain (Stich, 1998). Many philosophers think they are offering an analysis of *the* philosophically relevant concept such as knowledge, freedom, or intentional action. To give up the notion that we are analyzing *the* relevant concept may thereby seem to diminish the importance of some conceptual analyses. Concentrating on our own concept that may be shared by a few others may often be of relatively less interest than investigating a single

[7] Perhaps this is the point where our view diverges from Machery's (2017) view. At points, Machery points to being able to identify the validity and reliability of concepts. We think we can identify the reliability of intuitions—at least for some intuitions, the intuitions seem to be stably related to personality. We are less sure that, given our current techniques and practices, we are able to identify the validity of concepts (see discussion of Neo-Platonic projects). Perhaps we can, but this issue is not of central concern for our discussion of concepts since we only wanted to argue that intuitions could still be of use for conceptual analysis and that is consistent with Machery's view.

conceptual truth (Bishop & Trout, 2005). However, one benefit of the multiple concepts approach is that the variation of intuitions with personality does not pose a significant challenge to conceptual analysis and in some ways could help illuminate conceptual disagreement. Some people, because of their natural tendencies associated with personality (or other experiences and individual differences), may have different philosophically relevant intuitions. Indeed, as will become clearer in responses to objections to the PPA, it is important that the variation does not undermine some aspects of conceptual analysis. Consequently, the kinds of systematic differences in philosophical intuitions could be problematic for some approaches to conceptual analysis but not others.

Normative Projects

Normative projects are aimed at, among other things, telling us what we ought to do (ethics) and what we ought to believe (epistemology). What implications might the systematic differences in philosophical intuitions associated with personality have for normative projects? Somewhat of an aside, we want to note a striking lack of evidence about personality and epistemic intuitions. To our knowledge, there is no evidence suggesting that epistemic intuitions of the kind of interest to traditional epistemologists (e.g., about knowledge, justification, warrant) are robustly associated with personality. Of course, personality is associated with a host of different kinds of beliefs such as religious belief (Saroglou, 2002) and prejudice (Sibley & Duckitt, 2008), and personality disorders can be associated with distorted beliefs. But to our knowledge, research has not yet identified any data suggesting that personality is systematically associated with epistemic intuitions that are of interest to philosophers. Other researchers have reported individual differences in epistemic intuitions including gender (Buckwalter & Stich, 2014), culture (J. Weinberg et al., 2001), and socioeconomic status (J. Weinberg et al., 2001), but none of these studies reported personality differences (however, see Kim & Yuan (2015); Seyedsayamodst (2015) for failed attempts at replication).

One explanation for this lack of evidence is that there simply hasn't been enough research conducted on epistemic intuitions and personality. We take it that this is the most likely explanation since the research on personality and philosophical intuitions is relatively new. However, another explanation is that there simply is no interesting variation of epistemic intuitions and personality. One reason why there might be no interesting relation is that there might be something special about the nature of epistemic intuitions. For example, Jennifer Nagel (2012) has argued that many epistemic intuitions find their source in basic human capacities for mind reading. That is, most normally functioning humans have the basic ability to understand others' emotions, thoughts, intentions, and beliefs. Attributions of knowledge are like these other mind-reading abilities. Since we all have these basic abilities, then we all have the same epistemic intuitions, more or less. Consequently, we shouldn't expect there to be much variation with respect to epistemic intuitions and personality, gender, or socio-economic status.

Given the current state of the science, we are agnostic whether there are personality differences with respect to epistemic intuitions. However, we would like to note that if the reason why we shouldn't expect differences in epistemic intuitions is because of people's basic mind-reading abilities, then we should also not expect there to be differences in judgments of intentions among people with different personality traits. Attributing intentions to others is one of the paradigmatic instances of mind reading. However, we do see those differences in attributions of intentions between people with different personality traits (see Chap. 3). Whether similar differences can be found in epistemic intuitions remains to be seen.

Regardless of whether there are differences with respect to epistemic intuitions, there is growing and substantial evidence that there are systematic differences in moral intuitions associated with personality (see Chap. 4). There are already some well-developed theories concerning what implications systematic variation in moral intuitions could have for philosophical theories. As we have already discussed, Jess Prinz (2007) has argued that in order for somebody to make a moral judgment at all, one must have or have had an emotional reaction to the actions. If differences in personality cause differences in emotional reactions, then those differences could provide important pieces of evidence for normative views like Prinz's. Or, alternatively, Tamler Sommers (2012) has argued that the systematic diversity in intuitions about moral responsibility puts pressure on any universal, objective set of conditions for freedom and moral responsibility. If Sommers's is right, then those intuitions associated with personality could provide evidence for when somebody ought to be held morally responsible which could be different for different people with different psychological frameworks.

Not only does diversity of moral (and perhaps epistemic) intuitions help provide some evidence for some theories, the diversity can also help illuminate other, possibly surprising, debates (see Chap. 7 for a fuller discussion). For example, the diversity can help illuminate some contemporary debates about paternalism (e.g., Nudging (Sunstein & Thaler, 2003; R. Thaler & Sunstein, 2003; R. Thaler & C. Sunstein, 2008)). Given the systemic variation in some people's philosophical intuitions, it stands to reason that these differences sometimes are expressed and are related to people's values. However, paternalistic strategies, by their very nature, often aim to promote some specific set of values or goals. This means that if there is variation in people's values, then some people's values will not be served by the paternalistic strategy. Thwarting some people's values is an important cost associated with paternalistic strategies. If there are alternative ways to paternalistic policies to help people make better decisions, then all else equal those alternative ways should be evaluated to determine whether the paternalistic policy is more costly or if the alternative way of helping people is most costly. For example, *informing* people could allow for the benefits of paternalism (encourage better choices consistent with one's values) without commensurate costs (e.g., thwarting some sets of values).

Objections to the PPA

We now turn to objections that one might have to the PPA and its implications. While it would be impossible and inefficient to respond to every possible kind of defense one could mount against the implications of the PPA for intuitions in Neo-Platonic projects, in this section we focus on a cluster of common defenses. In evaluating the following objections, recall two plausible principles we have argued for earlier in the chapter:

> (E) Empirical evidence has a substantial role to play in many philosophical theories.
>
> And
>
> (R) Most fully adequate philosophical theories must take into account all available and relevant evidence

There are many ways one could reject (E) and respect (R). As we already indicated, whether using intuitions as evidence is appropriate depends a great deal on the goals one has using those intuitions. We think the strongest (and perhaps only) case for restrictionism is with respect to Neo-Platonic project, so we will use Neo-Platonic projects as our test case for objections.

The Self-Defeating Argument

Is the kind of view we are expressing even conceptually or theoretically possible? Some theorists think that it might not be. One potential way to show that our view is mistaken is by arguing that some of our assumptions call into question some other claims that we make. This general approach has been come to be known as the Self-Defeating Argument (Bealer, 1998; Bonjour, 1998; Goldman & Pust, 1998; Horvarth, 2010; Pust, 2001). In general terms, the Self-Defeating Argument holds that the only way to justify some of the claims we have made is to, well, use intuitions. But, if all those intuitions are associated with extraneous factors, then they should not be used for evidence (i.e., restrictionism about intuitions). So, on that argument, given our view, we might end up with nothing that could be used to support some of the key premises in the PPA. As Horvarth states, "the experimental challenge might easily lead to epistemic self-defeat because some of the relevant intuitions are themselves needed in order to justify the epistemic principles that form the very basis of the experimental restrictionists' own methodological criticism" (2010, p. 459).

In particular, restrictionists need evidence that the following principle is true:

> (EU+) If intuitions about hypothetical cases are unreliable and their unreliability is explicitly noted and explicitly noting their unreliability does not give rise to self-defeat, then they are not epistemically trustworthy. (Horvarth, 2010, p. 461)

Horvath observes that (EU+) is not "directly intuitive." That is, the intuitions are not directly generated by considering cases (similarly to the intuition A1, B2, and C2 combination above). Rather, the principle (EU+) involves testing out the principle on cases or other kinds of reflection. But, testing out intuitions in responses to cases or other kinds of reflection are exactly what the restrictionist aims to restrict—those intuitions are associated with extraneous factors that would call into question their contents. As a consequence, the restrictionist's view is self-defeating because the very evidence that is needed to support restrictionism is not available based on restrictionist principles.

Here it is important to mark again a distinction we (and others) have made. The PPA does not suggest that *all* philosophically relevant intuitions are suspect. We advocate "not the root and branch removal of all intuitions, but just the pruning away of some of the more poisoned branches" (Alexander & Weinberg, 2007, p. 71). Intuitions may be indispensable in projects in conceptual analysis and normative projects. Indeed, as we have seen, variability in intuitions related to personality can be theoretically important in those projects (cf., Prinz (2007)). The PPA supports restricting intuitions in some Neo-Platonic projects and also recommends scientifically documenting and tracking of non-trivial influences on intuitions for conceptual analysis and normative claims. As a result, there could be a wide swath of philosophically relevant intuitions that could be used to justify the premises of the PPA, (E), and (R). To illustrate, all the premises in the PPA are either descriptive or conceptual. (E) is a natural consequence of the PPA, and (R) seems to capture a straightforward conception of a fully adequate theory. We don't see any reason that we would be *required* to use Neo-Platonic intuitions to criticize Neo-Platonic intuitions (that would seem to be self-defeating). If we are not required to use Neo-Platonic intuitions and we do not use Neo-Platonic intuitions to justify the premises of the PPA, (E), or (R), then our argument supporting restrictionism is not obviously or necessarily self-defeating. Hence, it is not obvious or necessary that the PPA is self-defeating. Of course, one or more premises of the PPA could turn out to be false. But that criticism is very different from the criticism that somehow the PPA leads to epistemic self-defeat (and would require a different argument and evidence).

But perhaps the self-defeating argument can be cast in a slightly different way even if we restrict the range of intuitions to those used in Neo-Platonic projects. Regina Rini (2016) has offered a regress argument against those who claim that psychological evidence suggests that some moral judgments do not track the truth. Rini gives a schematic of the regress argument. The first premise states that some psychological processes give rise to some moral judgments. Examples of these kinds of psychological processes have been reviewed above and include order effects, mental correlates of socio-economic status, and personality. The second premise is that psychological processes do not reliably track the truth. Therefore, the moral judgment is unreliable. The problematic part of this schematic is that some moral judgment must be made about whether the psychological processes reliably track moral truth. For example, it is a *moral* judgment that order is irrelevant to the truth of moral claims. And, the argument goes, if one questions the psychological processes that give rise to some moral judgments, one must also question the psychological processes that give rise to other judgments. And the regress starts.

We think we are very clear that the PPA is problematic for Neo-Platonic projects and not necessarily for normative projects. As such, Rini's argument as stated does not pose any threat to the PPA. However, perhaps with slight modification, Rini's argument could generate a problematic regress. Here's how the modified schematic might look.

1. A set of Neo-Platonic judgments N are caused by psychological processes P.
2. P does not reliably track the truth
3. Therefore, N are not reliable.

But in order for us to determine the truth of (2), we have to know something about Neo-Platonic truths or have a set of Neo-Platonic beliefs on which to base our evaluation of (2). And if we question N, then why shouldn't we question the Neo-Platonic truths used to evaluate (2)? Hence, the regress starts.

We think a similar reply to Horvarth can be applied to Rini's argument. We need not appeal to any Neo-Platonic truths in the domain in question but rather to basic truths about what Neo-Platonic projects are conceived to be and some basic logic. Neo-Platonic projects attempt to understand the mind-independent, non-conceptual, non-linguistic understanding of some phenomenon. That means that Neo-Platonic attempt to find *the* truth about some phenomenon of philosophical importance. At this point, only the *concept* of Neo-Platonic projects has been employed, and not any Neo-Platonic claim about what Neo-Platonic projects actually are as mind-independent entities (if that even makes any sense). If that is granted, then it only takes some basic logic to show Neo-Platonic projects might not be able to be done if the intuitions are related to irrelevant features because two people could have contradictory intuitions about some philosophical phenomenon, and the content of both intuitions cannot be true. For example, if you have the intuition that *P* and I have the intuition that not *P*, they can't both be correct on a standard Neo-Platonic understanding. That would be like saying it is both true that the moon revolves around the earth and the moon does not revolve around the earth.[8] As a consequence, no regress is generated.

Defining Intuitions

Some may attempt to circumvent (E) and respect (R) by defining "intuition" so that intuitions cannot be empirically questioned. In this way, it can be claimed that the way restrictionists talk about intuition is "obviously not an intuition in the sense in which philosophers have talked about intuitions" (Ludwig, 2010, p. 437). For example, one may think that one has an intuition only when one has complete (enough) understanding of the relevant concepts involved and only makes judgments based on competencies with those concepts in ideal conditions (Bealer, 1998; Kauppinen, 2007; Ludwig, 2007, 2010). Indeed, some even claim that "it is impossible for intuitions properly understood to be relative" (Ludwig, 2010, p. 427) because there would be "identical judgments in response were the responses made solely on the basis of those competencies and identical understandings of the scenario, task, question, and adequate thought" (Ludwig, 2007, p. 145). As a result, intuitions are "veridical, and it follows that intuitions are not relative to cultures, socio-economic status, times, the ways questions are presented, or anything else, and this is demonstrable a priori" (Ludwig, 2010, p. 442). On this account of intuitions, when people have different responses to a scenario, at least one person does not have an intuition.

The variability in responses associated with personality is easy enough to accommodate while respecting (R) with an exclusive definition of intuition. Intuitions aren't relative to *anything*. However, the PPA suggests that intuitions are systematically related to one's personality. So, when people with different personalities give different responses, at least one of them would not have an intuition. Given that (R) holds that one needs to take into account available and relevant evidence, when there are divergent responses, at least one of those responses is not relevant because the response is not an intuition. As a result, these defenders of intuitions can easily deny (E) while at the same time as respecting (R).

Let's set aside the fact that there is substantial disagreement about what intuitions are (see above). Several theorists have argued that defining

[8] To be fair, Rini considers this as a possible, and legitimate, response to the "selective" debunking arguments. We simply make the response explicit here.

intuitions does not help the defenders of intuitions very much (see also, for a nice discussion, Weinberg and Alexander (2014)). First, *stipulating* a definition of intuition that makes intuitions invulnerable to empirical challenge is not very satisfying (Horvarth, 2010). But more problematically, intuitions defined in these ways may make it just as likely that philosophers don't have intuitions (Alexander & Weinberg, 2007). How could we ever tell when philosophers have an intuition? How do we determine (a) who the competent user of a concept is, (b) what ideal conditions are, or (c) if one's judgment is only influenced by semantic considerations (Kauppinen, 2007)? It seems like it would be very difficult, for example, to determine who are competent users or when ideal conditions obtain (A. Feltz, 2008; Kauppinen, 2007). To illustrate, Sosa writes that the reliability of intuitions "depend[s] on favorable circumstances in all sorts of ways, and these are often relevantly beyond our control. We must depend on a kind of epistemic luck" (2007a, p. 102). Restrictionists would argue that we are often epistemically unlucky when using intuitions in Neo-Platonic projects. There are hundreds if not thousands of studies across disciplines indicating that very minor changes in judgment environments (e.g., framing) can result in large differences in resulting judgments (e.g., Gigerenzer et al. (1999); D. Kahneman (2003)). But more to the point, it is very difficult to determine how lucky one is from the armchair. Horvarth (2010) notes there is quite a bit of evidence that we often aren't aware of all the causal influences on our responses to particular cases (e.g., Nisbett and Wilson (1977)). Given that we don't have much evidence that philosophers satisfy a–c, we should not be confident that philosophers even have intuitions. If philosophers don't have intuitions, then philosophers could not use intuitions in Neo-Platonic projects because they do not have any! We take it as part of the practice of philosophy that philosophers use the contents of their intuitions as evidence to do Neo-Platonic projects. So even if intuitions are philosophical theory mediated (e.g., one knows a lot about free will), the intuitions are used to provide the relevant contents, and it is hard to determine who, from the armchair, has contents reflective of genuine intuitions. Of course, we might be able to determine if philosophers satisfy (a)–(c). But since many factors relevant to determining if philosophers satisfy (a)–(c) are not introspectively discoverable, it won't be from the armchair. Hence, stipulating definitions of intuition does not insulate defenders of intuitions in Neo-Platonic projects from the PPA and (E).

The No Intuition Defense

Some defenses of philosophical practice attempt to show that the restrictionists argument is somehow misplaced because philosophy doesn't depend on intuitions at all. So, intuitions aren't used as evidence in philosophy, making premise 1 of the PPA false.

There are two major proponents of the no intuitions defense and they make slightly different arguments about why intuitions aren't used in any (substantial) sense in philosophy. First, we'll look at Hermann Cappelen's arguments. According to Cappelen, the key question is around *centrality* of intuitions in philosophical theorizing (i.e., Premise 1 of the PPA), or the claim that "contemporary analytic philosophers rely on intuitions as evidence (or as a source of evidence) for philosophical theories" (2012, p. 3). He goes on to argue that in no way is centrality a feature of philosophical debates.

According to Cappelen, it might look like philosophical debates involve something like centrality. One reason is that philosophers use "intuition talk." That is, philosophers often use phrases like "It is intuitive that x" that give the impression that intuitions are used as evidence. There are several problems with making the inference from intuition talk to intuitions actually being used as evidence for philosophical claims. First, there is no agreed upon definition of intuition, so inferring what the evidential role of something being intuitive is difficult, possibly highly context sensitive, and not automatically inferable from somebody saying that something is "intuitive." Additionally, there is almost nothing in philosophy (outside of some specific subdomains like logic) that is universally agreed upon to be intuitive. So not only are there problems with the definition of intuition, there are very few paradigmatic examples that are thought to be intuitive. For these reasons, when one encounters "intuition talk" in philosophical texts, one should be as charitable as one can be when interpreting what is being used as evidence. And, according to Cappelen, when one engages in charitable reconstruction, texts that contain intuition talk can almost always be recast without using intuition talk yet retaining the core of what the authors intend. This, in turn, suggests that intuitions don't serve as evidence at all because they are removable, can be interpreted as initial judgments subject to revision, or simply refer to pre-theoretical judgments or as initial starting points.

We agree that merely because philosophers use intuitions talk does not mean that they use intuitions as evidence. Premise 1 does not rely on intuition talk as evidence for its truth. Premise 1 requires that philosophers use the content of intuitions as evidence. So, the argument from intuition talk is no threat to Premise 1 of the PPA. Rather, a more important objection from Cappelen concerns whether intuitions are in fact used as evidence. According to Cappelen, there are three ways that would indicate that philosophers use intuitions as evidence.

1. Intuitions with a special phenomenology are used.
2. Intuitions that are rock bottom and in no further need of justification.
3. Intuitions that are the result solely of one's conceptual competence.

However, when one looks at the actual texts, there is no evidence of 1–3 in paradigmatic examples. For example, in Thomson's famous violinist case, one might think that the intuitions about the cases play some evidential role in the claim that a right to life does not always outweigh a right to determine what happens to one's own body. However, 1–3 are not present in the defense of that principle. According to Cappelen, Thomson does not refer to any special phenomenology when making judgments about the Violinist, the intuitions about the violinist are subject to scrutiny and are not taken to be rock bottom, and the intuitions are not solely the result of one's conceptual competence. In this paradigmatic case, therefore, intuitions do not play a substantial role. To the extent that intuitions do play a role, they play a role of setting a common ground from which to proceed (shared intuitions about the cases). As such, there is no evidence that intuitions play any substantial role in philosophy and if they do, their function is to provide a common framework and background in which to proceed philosophically. Consequently, Centrality is false and entails that Premise 1 of the PPA is also false.

Cappelen notes that Centrality (and by extension Premise 1 of the PPA) is an empirical claim about the actual way that philosophers go about philosophizing. And, as Cappelen notes, it's ironic that experimental philosophers take a remarkably non-experimental approach to substantiating that intuitions in fact play a role in philosophical debates. How many systematic studies have been done about intuition's actual role in philosophical theorizing? According to Cappelen, none.

However, it seems uncontroversial that philosophers make judgments about cases, principles, and premises in arguments. Of course, these judgments can be about a variety of things and can take into consideration a variety of different inputs. As already noted, philosophical intuitions are diverse and it is very difficult to identify exactly what an intuition is. Cappelen only identifies three of the many ways intuitions have been identified above. But for Premise 1 to be true, we don't need any well worked out account of what intuitions are. Rather, all we need is that, as a matter of fact, philosophers use the contents of intuitions as evidence—something that we do not take to be especially controversial or in need of any systematic studies. In fact, it is a virtue of Premise 1 that it does not take any substantive view about the nature of intuitions. All that is needed for the PPA to succeed is that philosophers have judgment (or intuition) biases and that only requires that the contents of intuitions (or judgments) are used as evidence for philosophical claims. Looking again at Thomson's violinist example, it seems uncontroversial that we have some intuitions about the case and the contents of those intuitions are supposed to carry evidential weight. Of course, we can examine these contents, refine intuitions, and reject the content of some intuitions. However, in each of these instances, the contents of the intuitions are used as evidence for some philosophical claims. Hence, Premise 1 of the PPA is not shown to be false by Cappelen's argument.

Max Deutsch (2015) has a slightly different approach to the no intuition defense. Deutsch agrees that if intuitions do play an evidential role in philosophy, then the results like those associated with personality would be deeply problematic for philosophy. However, the essential feature of philosophical debates is arguments and not intuitions. If intuitions don't play an evidential role in philosophy, then Premise 1 of the PPA is false.

Deutsch laudably makes the distinction that we have exploited between the mental state of having an intuition and the content of the intuition. In terms of this distinction, one might think that philosophical evidence comes from something's intuitiveness (EC1). For example, the intuitiveness of the Gettier examples is evidence against the justified true belief account of knowledge. However, on a very different notion of intuition (EC2), it's the content of the intuition, not the fact that one has the intuition that is evidence for its philosophical claims. So, it's not that the intuitiveness of the Gettier cases that is evidence. Rather, it's the content of the intuition—that the person doesn't know—that is evidence against the JTB account. Deutsch contends that the EC2 is without question true but not EC1. And many experimental philosophers mistakenly take that EC1 is the way that intuitions are evidence for philosophical claims.

Critical to Deutsch's claim is that EC1 doesn't at all play a role in philosophy, which we are happy to grant. Additionally, good use of intuitions (characterized by EC2) are backed up by arguments. We are happy to grant that as well. In this way, it's not the case that Deutsch simply relocates the problem to a different set of intuitions that are used to justify the premises in arguments. All intuitions could be questioned and arguments could be provided for them (e.g., some form of coherentism). As such, one might think that either Premise 1 of the PPA is false (intuitions aren't used as evidence) or think that the variability we find is not problematic because arguments could be given for the different positions, and the best arguments will win out.

Here, it becomes useful to be reminded of the distinction that has been drawn between Neo-Platonic projects, projects in conceptual analysis, and normative projects. We agree that the right way to understand intuitions is not simply that something is intuitive (EC 1). Rather, the content of the intuition is what is critically important for philosophical practice (EC 2) (e.g., used as evidence). The problem is that people have intuitions with different contents, or those contents occur with different strengths. Take, for example, the debate about manipulation in the free will debate. Some argue that moment-by-moment manipulation rules out freedom and moral responsibility. The notion is that people have the intuition with the content that one cannot be free and morally responsible if all of one's actions are ultimately caused and implanted by somebody else. However, hard-liners don't have this intuition or at least they don't have that intuition about some cases (e.g., McKenna (2008)). Rather, hard-liners have the intuition that if the person appropriately identifies with the actions, that person is free and morally responsible even if manipulated. Here, we don't see a clash of intuitiveness (EC1), we see a clash of the contents of intuitions where some people have the intuition that manipulation always rules out freedom and moral responsibility whereas others have the intuition that manipulation does not always rule out freedom and moral responsibility (although we are also likely to see clashes of intuitiveness (EC1) as well). Arguments are given by both sides, but still in many cases, the intuitions are unwavering. Moreover, these intuitions are predicted by personality traits (see Chap. 2). So even though hard-liners and non-hard-liners are possibly justified in having those intuitions because they both have coherent sets of intuitions and arguments, the coherent sets are different. As such, it cannot be the case that both sets accurately capture the relation of manipulation to free will as it exists in a Neo-Platonic sense. To be sure, different people may have accurately captured their *concept* of freedom, and the differences may have important *normative* work to do, but they do not help us solve *Neo-Platonic* questions. So, even if everything that Deutsch says is correct, the PPA still can have problematic implications for Neo-Platonic projects.

At this point, one may object that it is not clear how extensive this kind of philosophical disagreement is. Perhaps debates around manipulation are rare, non-representative cases of philosophical disagreement. In other cases of philosophical disagreement, perhaps extensive philosophical reflection can help achieve convergence. If convergence is not achieved, perhaps this extensive reflection can at least reduce the influence of non-truth tracking features. So, how extensive and robust is the influence of non-truth tracking features? Answering this question is an empirical question and not purely a philosophical or theoretical question that can be handled from the armchair. Here, again, we want to underscore the fact that we think that given the available evidence many Neo-Platonic *risk* not being able to be done. However, before we can be sure that this actually is the case we need to have more evidence. Given the available evidence, our bet is that philosophical disagreements about Neo-Platonic projects are deep and are not likely to go away soon—and if they do, it's not going to be because of arguments that justify some intuitions over others.

The Verbal Defense

According to the Verbal Defense, the current evidence based on surveys does not ensure "true disagreement" in people's intuitions (Horvarth, 2015; Sosa, 2007a). In order for there to be true disagreement, responses gathered by experimental philosophers must be about the same things. But the worry is that different people could interpret scenarios or questions differently and thereby have intuitions in response to different things. There are a number of ways that people could interpret scenarios differently. To illustrate, Sosa (2009) argues that the materials many experimental philosophers use are like stories. Like most stories in fiction, not all details are spelled out in the text. As a result, people often fill in stories in different ways. Participants may do the same thing for the scenarios used in experimental philosophy. People simply fill in the scenarios differently and are thereby representing relevant content of the scenarios differently. These differences may result in different intuitions but not about the same things. Likewise, people may interpret questions asked somewhat differently. For example, when asking whether somebody is morally responsible for an action, people may interpret "morally responsible" in a variety of ways. They may interpret moral responsibility in an attributability sense where judgments are made about the action reflecting the actor's character. Or, participants may interpret moral responsibility in an accountability sense where one can be held accountable (e.g., punished/rewarded) for acting (Sosa, 2007a). If participants interpret scenarios differently or interpret questions differently, then much of the disagreement in intuitions put forward by experimental philosophers is merely surface or verbal disagreement. In the end, people could be "talking past each other" (Kauppinen, 2007, p. 107). Since such surface variability is not philosophically relevant, we can reject (E) while respecting (R).

The PPA may seem to support the verbal defense. As we have documented, people with different personality types have different sensitivities, beliefs, and goals (Costa & Mccrae, 1988; Funder, 2001). These differences may result in people with different personalities resolving ambiguities in scenarios differently.[9] So, it might be that people with different personality types are not disagreeing about the same contents of the intuitions.

However, if the PPA makes it plausible that apparent disagreement among the folk is only verbal disagreement, it makes it just as likely that disagreement (or agreement) among philosophers is not disagreement (or agreement) about the same things (Alexander & Weinberg, 2007). That is, we cannot be sure that agreement or disagreement among philosophers is about the same content of the intuitions. Not being able to tell when there is true agreement and disagreement results in general skepticism about philosophical uses of intuitions because we could never tell when philosophers are truly agreeing or disagreeing (Alexander & Weinberg, 2007; Machery et al., 2004). Hence, if the verbal defense were to succeed, it would be at the expense of skepticism about intuitions in general. But we take it that philosophers (experimental or otherwise) want to resist general skepticism about intuitions.

[9] There is some evidence that some people may disambiguate scenarios differently, at least in free will. For example, some people, but not others, may import non-determinism into deterministic scenarios when making free will judgments (Nadelhoffer, Rose, Buckwalter, & Nichols, 2020a). For similar differences associated with general cognitive abilities, see Feltz et al. (2022).

Just as in the Expertise Defense, one may argue that the burden of proof may be on the experimentalist to show that philosophers run the risk of only having verbal disagreements. "The appeal to divergence of interpretation is a *defensive* move, made against those who claim that there *is* serious disagreement in supposed intuitions. It is only against such a claim of disagreement that we must appeal to verbal divergence. But any such claim need be taken seriously only when adequately backed by evidence" (Sosa, 2007a, p. 103). Fair enough. But there is evidence emerging that adequately backs the worry. As already noted, experts about the free will debate are influenced by specific, heritable facets of their general personality traits: Warm extraverts tend to have more compatibilist friendly intuitions than do introverts (Schulz et al., 2011). As we understand Sosa's defense, any purported disagreement in intuitions is verbal and that would entail that when there is disagreement between experts, that disagreement would largely be a verbal disagreement. That would mean that the disagreement in expert intuitions between non-warm extraverts and warm extraverts runs the risk of being merely verbal. If that is the case and the empirical results generalize, the verbal defense runs the risk of it being hard to tell when there ever is a substantive disagreement among philosophical experts such as in free will. Consequently, the verbal defense along with relevant empirical evidence risks general skepticism about philosophically relevant intuitions.

Intuition Calibration

The last objection we will consider is what might be called the intuitions calibration objection (for a more extensive discussion of intuition and calibration, see Weinberg, Crowley, Gonnerman, Vandewalker, & Swain, 2012b). Rather than objecting to the relevance of the variance associated with personality (e.g., arguing that philosophers, for whatever reason, won't display that bias), the intuitions calibration approach starts by accepting the basic empirical science. That is, this objection freely grants that philosophers often display the same (or similarly problematic) biases as non-philosophers. However, philosophers don't *have to* display that bias and can in fact incorporate those biases in their theorizing. One way that might happen is by attempting to correct for those biases in intuitions. In this sense, philosophers correcting for a bias in their intuitions are much like other debiasing strategies that are often successful. Take one example: the Planning Fallacy. The planning fallacy is the tendency for people to underestimate the amount of time, effort, or other expenses that projects can require. You may have fallen prey to the planning fallacy yourself when you, for example, thought about how long it would take to write a paper or clean your garage (or write a book!). However, you might also know some of the psychology about the planning fallacy and incorporate that into your estimates of how long it would take to do something. Perhaps you double the amount of time you estimate and that ends up being roughly accurate to the amount of time it actually does take. By incorporating that information, you have removed the bias. Philosophers might be able to do something similar with personality effects—they could use that information to calibrate their intuitions more accurately.

While we agree that debiasing approaches often work very well, we are less convinced that they are going to be of much help to correct for philosophical biases associated with personality. Recall we are arguing a restricted restrictionist's view that intuitions are not likely to be reliable guides truth for Neo-Platonic projects. These projects attempt to discover the truth about some phenomenon, independent of what anyone thinks of those truths, using intuitions as indispensable sources of evidence (see Cummins (1998) for a discussion related to calibration issues in philosophy). These intuitions are at least in part used to determine those neo-Platonic truths and therein lies the potential problem with the calibration approach. For most (if not all) of the standard cases where debiasing happens, there is some external standard that is used to determine if one's judgments, decisions, or estimations are accurate. In the Planning Fallacy example above, there is some external, objective measure of how long a project actually takes—time. One's intuition or estimation about how long a project takes does not factor at all into the measure of how long the project actually takes. But for many philosophical areas that attempt to discover Neo-Platonic truths, the intuitions constitute or are a guide to what those truths are (see Sommers (2010); for a dissenting view, see Timothy Williamson (2007)). Given that intuitions play this role in many Neo-Platonic projects, and given the empirical data that often those intuitions are related to extraneous features like personality, there is no external standard on which to calibrate those intuitions. This variation seems to persist, at least sometimes, in the face of extensive knowledge and training in a subject area. Finally, we know of no argument that allows us to prefer some intuitions over others (should we prefer extraverts intuitions? If so, why, and does that preference require further intuitions?). Consequently, addressing the calibration problem does not, at least at present, insulate Neo-Platonic projects from the risks that we have identified.

Concluding Remarks

If the PPA is sound, some deep-seated ancient debates in some Neo-Platonic projects may not go away without new methods. The reason for this may be surprising. There is some evidence that global personality traits are at least partially genetic in origin (Bouchard & Loehlin, 2001; Jang et al., 2002). And, as noted, intuitions are used extensively in many philosophical projects. Thus, some intuitions used in Neo-Platonic projects are systematically related to personality and personality is at least partly (and often largely) heritable. If both of these two claims are true, then one's tendency to endorse a particular philosophical view is also at least partially inherited. So, some issues in Neo-Platonic projects are likely to persist through the generations of philosophers with very little resolution. One could take this result as pessimistic, but we think it is actually encouraging. It may help free philosophers from an over-reliance on intuitions and may help encourage philosophers to use other methods and evidence to cover new ground for important, ancient Neo-Platonic projects.

Our view that is a consequence of the PPA might also have implications for inclusivity and diversity in philosophy. Systematic diversity associated with personality is at the core of our argument for the PPA. Perhaps this recognition of systematic diversity may help temper and inform philosophical debates (even if it doesn't necessarily help us calibrate those intuitions). Philosophical debates are about important issues and many people feel deeply about those issues. Sometimes when there is philosophical disagreement, parties to the disagreement think that the other person is making some sort of mistake. That mistake can be of a variety of forms including having false beliefs or lack of relevant beliefs or not thinking appropriately. However, the evidence we have reviewed here suggests that there may be something else going on. Rather than attribute some epistemic deficit to an interlocuter, that interlocuter may simply have different intuitions (and possibly concepts) involved in the dispute. This recognition might allow different ways to resolve some disagreement and to explore the sources and implications of that diversity (for similar reasoning, see Haidt and Graham (2007)).

At this point, we have reached the end of our empirical defense of the premises in the PPA. Before we turn to some of the more practical implications that the PPA might have, let's take stock. We think we have established that the PPA is likely to be sound. The soundness of the PPA suggests that at least for some philosophical projects, restrictionism is warranted. If one wants to deny restrictionism, then one would have to provide evidence that somehow the PPA is not valid (it seems to be of a valid form, however) or provide evidence that the PPA has at least one false premise. As an empirical matter, one or more of the premises in the PPA maybe turn out to be false. We think that the evidence reviewed so far makes this possibility very unlikely. Nevertheless, the kind of evidence that would be needed to falsify one of the premises of the PPA *is empirical evidence*. While we think that our view supports restrictionism, we would be very satisfied if we have achieved the conclusion that at minimum philosophy should be substantially more empirical (Alexander & Weinberg, 2007; Feltz & Cokely, 2012a, 2013a; Feltz & Cokely, 2016).

CHAPTER 7

Ethical Interaction Theory

We started this book with a simple ethical question: *How should you live your life?* In closing, we turn to a related question: *Who should decide how you live your life?* This new question may seem strange given where started, but it's important because we have recently witnessed the emergence of a powerful behavioral targeting revolution. This modern technology-driven reality is capable of shaping, tracking, and controlling many of our behaviors and decisions, while simultaneously influencing much of what we are exposed to (e.g., digital marketing, social media content, personalized browsing). Some related behavioral science advances are being leveraged by governments to promote prosperity and innovation across public and private sectors (e.g., U.S. Presidential Executive Order—*Using Behavioral Science Insights to Better Serve the American People*). Unfortunately, there is also an inestimable potential threat resulting from the rapid rise of science and technology that efficiently manipulates perceptions and decisions without any consent or awareness of those manipulated. Even for some of the most popular scientific frameworks that have been explicitly designed to help people make decisions in their own best interests (R. H. Thaler & C. R. Sunstein, 2008), some common applications appear to be just as ethically questionable as some predatory profiteering schemes or social credit systems. In the light of these rapidly evolving and powerful technologies, in what follows we present a novel conceptual framework to inform the design and evaluation of choice architectures and interactive systems. Specifically, based on the evidence and theory presented in previous six chapters, in this chapter we will establish a foundation for practical

A. Feltz, E. T. Cokely, *Diversity and Disagreement*,
https://doi.org/10.1007/978-3-031-61935-9_7

guidance on methods and heuristics for *Interactive Policy Analysis* (e.g., policy design, evaluation processes, and standards) based on *Ethical Interaction Theory*—i.e., a normative theory that provides a philosophically grounded and evidence-based account of how, why, when, and for whom various interaction policies and choice architecture are likely to be more or less ethical and efficient.

It is probably not surprising that Ethical Interaction Theory incorporates the diversity of fundamental philosophical intuitions we have reviewed in the previous chapters. But Ethical Interaction Theory also turns on one additional basic assumption, roughly stated as follows: *All else equal, we assume that every person who is competent should have the opportunity to make their own decisions in most situations, absent unwanted infringement on the autonomy of others* (e.g., *so long as it doesn't hurt other people*). While this is a broad assumption with many nuances and caveats, it is also a relatively uncontroversial assumption that has been codified in wide-ranging practices and widely accepted standards for informed decision making (e.g., bioethics, law, professional ethics codes; Benn, 1976; Drane, 1985; Dworkin, 1981, 1988; Felsen, Castelo, & Reiner, 2013; Haworth, 1986; Mele, 1995).

To ensure Ethical Interaction Theory is philosophically and empirically sound, in the first half of this chapter we map connections among diversity of philosophical intuitions (Chaps. 2–6), scientific findings, and human values. We then consider theory and science related to one of our mostly widely agreed-upon values—i.e., *autonomy*—and its role in informed decision making and human well-being more generally. This is followed with a review of some powerful emerging approaches to non-rational persuasion techniques used to shape decision making without technically limiting choice—i.e., *Libertarian Paternalism*. Ultimately, we show that the fundamental philosophical biases and disagreements associated with personality that we have extensively documented in this book profoundly complicate arguments supporting Libertarian Paternalism. These complications are especially pronounced under conditions in which there are other known and available strategies that promote ethical interactive systems (e.g., transparent decision aids and representative educational materials). We then argue that the weight and breadth of the evidence implies that informed decision making is generally ethically and practically superior compared to alternative non-rational persuasion and paternalistic policies, especially in the context of value diversity related to fundamental philosophical values. However, we want to be very clear that Ethical Interaction Theory *does not* imply that one type of choice architecture or interactive policy is *always* better than another. In contrast, the framework simply holds that although some factors are ethically and practically preferable, this superiority may only be obtained under specific conditions. As such, under other common conditions viable alternatives (e.g., Nudges) may become preferable given necessary trade-offs among essential values and ethical priorities including autonomy, efficiency, and beneficence (e.g., sometimes it is too expensive or too slow to inform everyone).

Ultimately, it is not primarily the *outcome* of a decision that is the target of our ethical interactive policy analysis (e.g., the decision need not capture some Neo-Platonic truth about correct answers). Rather, what is ethically evaluated is the policy that determines the *process* that is used to shape how people interact with systems (or with each other), thereby shaping their judgments and decisions. In other words, interactive policy analysis involves strategies, tools, and methods that may be used to help evaluate ethical threats and vulnerabilities in systems designed to shape interactions (i.e., the *how to* of analysis), whereas Ethical Interaction Theory provides the theoretical foundation and philosophical justification for such analyses (i.e., the *why* and *when* of analysis). To begin to make our case, we turn our attention first to conceptions of human values and their connection with philosophical intuitions.

Values, Philosophical Intuitions, and Personality

What are "values" and how should we characterize the relations between values and fundamental philosophical intuitions? Clearly, one reason theorists purport to study basic philosophical issues is that these issues have some important implications for health, wealth, happiness, justice, and so many of the other things people value deeply (Bishop, 2015; Bishop & Trout, 2005; Kane, 1996). Indeed, fairly uncontroversial and converging conceptions of what (human) values are have been addressed across academic disciplines, including philosophy and behavioral science. For example, values are often said to be anything deemed good or that are appropriate to desire (Velleman, 2008). For practical purposes, we find the influential account offered by Ruth Chang instructive, wherein "A 'value' is any consideration with respect to which a meaningful evaluative comparison can be made" (Chang, 1997, p. 5). More specifically she notes:

> [Values] can be oriented toward the good, like generosity and kindness; toward the bad, like dishonor and cruelty; general, like prudence and moral goodness; specific like tawdriness and pleasingness-to-my-grandmother;

> intrinsic, like pleasurableness and happiness; instrumental, like efficiency; consequentialist, like pleasurableness of outcome; deontological, like fulfillment of one's obligations; moral, like courage; prudential, like foresight; aesthetic, like beauty; and so on. (Chang, 1997, p. 5)

Some of the items on this list might not initially strike you as values, such as fulfilling one's obligation. But, on Chang's view, we have clear and relatively non-controversial standards that allow us to unequivocally characterize them as such. That is, since we can *compare and evaluate actions* with respect to whether the action fulfills an obligation, fulfilling one's obligation can and should be considered a value (similar conceptions can be found in the behavioral sciences, e.g., Schwartz and Bilsky (1987)).

Given standard and well-accepted notions of values, tight and manifold connections between values and the kinds of philosophical intuitions that are linked to personality that we've discussed throughout this book become clear. Consider some obvious cases in ethics. Some philosophers think that objectively wrong or right actions carry with them a special status compared to things that are judged to be conventionally right or wrong. We can expect that many people think that eating one's soup with one's salad fork is wrong but not as wrong as eating one's neighbor with one's salad fork. The latter may be perceived as objectively wrong, whereas the former may be perceived as conventionally wrong. And, the tendency to judge things objectively wrong as worse than things that are conventionally wrong is one way people can make better, worse, or equivalent to judgments. As such, this is an instance wherein one's intuitions about moral objectivism inform a judgment of whether some actions are morally better, worse, or equivalent to others. Thus, ethical intuitions can directly shape and inform some values. Additionally, the tendency to have intuitions consistent with moral objectivism is diverse and linked to personality traits (e.g., openness to experience).

Intuitions about free will and moral responsibility also reflect and involve values in many obvious ways. If a person is morally responsible for an action, then that person is a more apt target for praise and blame than a person who is not. If there are degrees of moral responsibility (Mele, 2008), then judging that someone is more morally responsible should factor into judgments of more praise or blame for those actions compared to actions for which one is less morally responsible. As we have documented, the tendency to judge a person free and morally responsible in a variety of different contexts is related to personality traits (e.g., extraversion) suggesting that people have diverse values concerning moral responsibility.

Of note, there is a non-accidental connection between autonomy and freedom. Sometimes theorists simply take freedom and autonomy to be synonymous (for a discussion see Dworkin, 1981). From these perspectives, if autonomy is valuable then so is freedom (Haworth, 1986). And, it is almost universally agreed upon that autonomy is a value (cf. Skinner (1971) and see next section for a related review). Moreover, some argue that freedom underwrites things that we value such as friendship, worth of actions, and so much more (Kane, 1996). On these views, freedom has at least some instrumental value and can profoundly shape values more generally.

Intentional action intuitions also involve and reflect values in many ways. Like judgments of freedom and moral responsibility, intentionality judgments are important elements in how much praise or blame we attribute to somebody. This fact is reflected in everyday judgments where we blame somebody more for actions done intentionally compared to unintentionally (e.g., stepping on your foot). The values associated with intentional action are also reflected in ubiquitous legal standards throughout industrialized countries where typically the most severe punishments are reserved for somebody performing an action intentionally. Because intentionality judgments can be used in comparisons such as these, they can involve, reflect, or be values. Given the frequency with which people appear to reflect on their own intentions as well as those of others, here too it seems obvious that these intentional action intuitions may often be values in many contexts. Again, just as was the case for ethical and free will intuitions, some intentional action intuitions are associated with personality traits (e.g., extraversion) suggesting that there are stable, yet diverse values concerning intentional action.

In the light of these and many other examples, the weight of the evidence suggests that by and large philosophically relevant intuitions commonly reflect or are reflections of our values, and are often associated with personality. Of course, this does not imply that every single philosophical intuition or belief is related in some way to values or personality. Rather, it is sufficient for our current analysis that some philosophical intuitions are connected to values and that these philosophical intuitions are predictable, diverse, and stable. As we have argued, the diversity of philosophical intuitions associated with personality along with the conclusion of the Philosophical Personality Argument means that we may not be able to do some Neo-Platonic projects. The inability to do Neo-Platonic projects poses special challenges to common paternalistic strategies. In short, we may not be able to identify *the single value* to promote with the paternalistic policies and that inability may have important ethical costs that should be evaluated.

Autonomy: One Shared Value

As we have argued, many values and philosophical intuitions appear to be diverse and related to personality. However, there is surprisingly wide and enduring agreement on others. This fact does not entail that these values are right (in a Neo-Platonic sense). Nevertheless, the convergence and acceptance of some values are practically and theoretically noteworthy. Out of the many seemingly shared values, one such fundamental and widely shared human value is *autonomy*. While any well-specified definition of autonomy is philosophically contentious, all accounts in some way capture the central notion that autonomy involves people being self-determined and making informed decisions in accordance with their values (Benn, 1976; A. E. Buchanan & Brock, 1989; Dworkin, 1981, 1988; Ellis, 2008; Mele, 1995).

Many accounts of autonomy converge that the value of autonomy can be either instrumental (i.e., helps bring about other things that have value) or intrinsic (i.e., valuable in and of itself). Our review of the literature reveals great consistently across philosophical and empirical accounts on the instrumental value of autonomy. For example, Bentham (2008) famously noted that autonomy is an instrumental good that leads to higher overall well-being, which is a finding that is today among the most well-established in the scientific literature on psychological health, well-being, and achievement (e.g., self-determination theory; Deci & Ryan, 1995; see also Bandura, 1986; Peterson & Seligman, 2004; Seligman & Csikszentmihayli, 2000). Likewise, John Stuart Mill captured this sentiment when we wrote: "If a person possesses of any tolerable amount of common sense and experience, his own mode of laying out his existence is the best, not because it is the best in itself, but because it is his own mode" (Mill & Williams, 1993, p. 135). In these ways and others, philosophical accounts and empirical analyses demonstrate how and why autonomy contributes to human health and welfare and may generally support some other, perhaps more basic, values (e.g., justice, well-being) (Dworkin, 1988). The instrumental value of autonomy may also in part explain why so many practical policies and protections for autonomy have been institutionalized throughout modern societies and organizations (e.g., in professional ethics codes).

Clearly, the instrumental value of autonomy is well-established, but for many autonomy is also an intrinsic value. On these views, autonomy is good not (only) for anything that autonomy helps bring about, but it is good just for its own sake. For example, some people value making their own decisions that are expressions of their values and desires. They may value making these kinds of decisions for their own sake and for no other reason. For better or worse, we see that many people want to have the freedom to make independent, even if potentially poor, choices. The desire for the freedom to make potentially poor choices is clearly not because the ability always (or even on average) brings about good consequences (i.e., the bad choice actually brings about bad things, on average). Rather, the value is because it was "I" who made that choice. Theoretically, that state of affairs could have value in and of itself independent of any outcomes. As such, the presence of environmental constraints or even beneficial manipulators can infringe on this fundamental value because it is not "I" who is the (primary) author of change (Benn, 1976). Ultimately, for many it is simply important to be the authors of our own lives (Dworkin, 1988). Whether autonomy is an instrumental or intrinsic value (or both), autonomy is commonly thought to have significant value—a notion that is widely endorsed by experts and folk alike. But autonomy is not the only value, and under the right conditions, it is widely agreed that autonomy should be violated. In the next section, we discus some of the instances when autonomy can be violated and why.

The philosophical work on the value of autonomy is also reflected in empirical work and scientific theory. Schwartz and Bilksy (1987) conducted extensive cross-culture studies and they arrived at converging conceptions of values based on empirical studies of diverse people (and professionals). They found eight basic values that appeared to be universal in humans, which were later revised to ten basic values in the light of more comprehensive data and analyses (Schwartz, 1992). Among these values was the value of being self-directed. And on all accounts of personal autonomy that we are aware, self-direction is a core element. Consequently, not only is autonomy thought to be important philosophically there is good evidence that autonomy is in fact valued by people in various cultures.

A short set of principles provides an efficient basis for a discussion of how autonomy may factor into personal choice and broader policy debates.

For example, consider Mele's set of jointly sufficient (but not necessary) conditions for autonomy. According to Mele, a self-controlled[1] person acts autonomously if:

1. The agent has no compelled motivational states, or any coercively produced motivational states.
2. The agent's beliefs are conducive to informed deliberation about all matters that concern him.
3. The agent is a reliable deliberator. (A. R. Mele, 2001)

The first condition states that one's motivation should not be the result of things such as uncontrollable phobias or brainwashing. Mele's second condition holds that one's beliefs should not be the result of false or misleading information on which the person deliberates before deciding. The final condition covers the skills, conditions, and habits used to deliberate effectively about means and ends.

One set of factors Mele identifies is particularly important for our purposes—namely, being informed and competent (conditions 2 and 3). These two factors are constitutive of what we will call *rational agency*. Rational agency characterizes the state where one is competent and informed and can integrate information and one's values into decisions. The relevant decision making pathway is consciously accessible and the person is actively involved in the decision. Broadly, in accord with standard philosophical conventions, we take rational agency to refer to a set of capacities (i.e., competence and being informed) that allows one to take information and representatively and coherently integrate the available information, values, and prior beliefs to make a decision (J. Baron, 2008; Weirich, 2004).

However, there are many influences on people's decision making, some of which do not factor into rational agency. As detailed in Chap. 5, the underlying psychological processes involved in boundedly rational agency don't require that a decision maker be neo-classically rational or employ formal normative decision analyses during decision making (e.g., deriving and solving a statistical equation in one's mind or with the help of a computer). On our scientifically informed view, adaptive (boundedly) rational agency generally only requires that decision makers *use* the relevant

[1] The qualification of the self-controlled person is important to rule out cases of weakness of will.

information, along with their relevant values, to reach a locally coherent *representative understanding* of a decision that robustly accords with standards of normatively superior decision making (e.g., aligns with but does necessarily follow from logical, probabilistic, and statistical standards; see Skilled Decision Theory, Cokely et al., 2018; see also Gerd Gigerenzer et al. (1999); Levi (1967)). Of course, because people are not logical super-computers, sometimes the way that information is presented will predictably bias even the most skilled and informed decision makers.

To illustrate, people can be persuaded, coerced, or influenced by a number of (potentially) non-rational factors like the way that information is framed. Framing can happen when essentially logically identical, but different, descriptions of a choice are used to structure the presentation of information (for a review, see Levin, Schneider, and Gaeth (1998)). Among several robust behavioral biases that can result from framing, one influential bias exhibited by many people is Loss Aversion: People act as if losses loom larger than equivalent gains (almost three times larger on average). As a result, even when presenting people with logically equivalent information, people's choices can be biased by framing choices with respect to the potential gains versus losses involved (see Chap. 5 for other examples).

Paternalism and Nudging: Features of Some Choice Architectures

There is no question that people often make bad decisions. They decide to do some things that are not in line with their own best interests or their own values. Sometimes, these decisions are driven primarily by environmental factors (e.g., framing, time constraints, ignorance). It may be reasonable to assume that in instances where people make predictably bad decisions it is ethically justified to intervene on their decision making to encourage those people to make better decisions. But the question is *how best* to intervene?

One way to intervene on decision making is to adopt some paternalistic policy. Paternalism, like most philosophically complicated concepts, is somewhat difficult to define precisely and satisfactorily. There is no consensus on any single definition of paternalism (Trout, 2005). Some think that the core element of paternalism is a violation of a person's autonomy (Dworkin, 1988). Others think that one of the essential features of paternalism is the willful withholding of important information or the providing of false or misleading information to decision makers (A. Buchanan, 1978). For most practical purposes, Gert and Culver's analyses of paternalism is instructive and representative:

> A is acting paternalistically toward S if and only if A's behavior (correctly) indicates that A believes that (1) his action is for S's good; (2) he is qualified to act on S's behalf; (3) his action involves violating a moral rule (or will require him to do so) with regard to S; (4) S's good justifies him in acting on S's behalf independently of S's past, present, or immediately forthcoming (free, informed) consent; and (5) S believes (perhaps falsely) that he (S) generally knows what is for his own good. (Gert & Culver, 1979, p. 199)[2]

Accordingly, the justification for any paternalistic policy is that the overall benefits that are accrued by the policy outweigh costs associated with the policy. The benefits are supposed to be for the individual for whom the policy is designed. However, these benefits come at the cost of violating some moral rule or violating some other moral good. For example, seat belt laws are often thought to be justifiable paternalistic policies. The benefit to individuals (reduction of risk of death and injury) justifies the infringement on personal freedom (which is typically thought to be a significant moral cost). Even though it is possible that nobody ever consents to the seat belt laws, the policy is justified based on the overall reduced risk of death and injury.

Nudging is one popular recent strategy that may greatly reduce the moral costs associated with "hard" paternalistic policies like seat belt laws (Johnson et al., 2012; Oliver, 2015; Sunstein & Thaler, 2003; R. Thaler & Sunstein, 2003; R. H. Thaler & C. R. Sunstein, 2008). According to R. H. Thaler and C. R. Sunstein (2008), nudging is

> [A]ny aspect of the choice architecture that alters people's behavior in a predictable way without forbidding any options or significantly changing their economic incentives. (p. 6).

To be clear, nudges are just one specific type of Choice Architecture. Choice Architecture more generally has been described as "The idea that changes to the decision environment can affect individual decision making

[2] See also Feinberg (1986).

and behavior" (Muenscher, Vetter, & Scheuerle, 2016).[3] The idea that we can and should scientifically design *interfaces* and *affordances* (e.g., choice architectures) to facilitate and enhance human and sociotechnical system performance has a nearly century-long history in scientific sub-disciplines such as human factors, cognitive engineering, and engineering psychology (Wickens, Hollands, Banbury, & Parasuraman, 2013), in addition to many other historically long-standing efforts in marketing, business, political science, public relations, and communications (e.g., propaganda). What is somewhat more novel in recent years is the widespread, intentional application of scientifically grounded efforts applied for wider *benefit* to the decision maker themselves, via public or institutional policies.

Consistent with notions of paternalism and psychological biases, it is commonly thought that nudges "are called for because of the flaws in individual decision making, and work by making use of those flaws" (Hausman & Welch, 2010, p. 136). These flaws are often said to be the result of automatic processing that the nudges can take advantage of (Hansen & Jespersen, 2013). For example, Johnson et al. (2012) state "the same factors that lead us to make a mindless suboptimal or unhealthy choice can often be reversed to help us make a mindless better choice" (p. 500). Selinger and Whyte (2011) think that the characteristic feature of a nudge is that it changes the context "in subtle ways that often function below the level of our conscious awareness, to make decisions that leave us and our society better off" (p. 925). Muenscher et al. (2016) say that nudges "can be understood as a specific type of behavior technique primarily relying on reflexive cognitive processes." Across these and many other characterizations, the common reflexive cognitive processes that are usually leveraged by nudges are not part of one's conscious, rational agency (even if the automaticity is adaptive), and therefore nudges often bypass one's conscious rational agency.[4]

On our understanding, some such nudges may qualify as instances of an approach to paternalistic policies called Libertarian Paternalism. Libertarian

[3] See also R. H. Thaler, Sunstein, and Balz (2012).

[4] Some researchers have defined nudges in such a way that any intervention that improves behavior is a nudge (e.g., education, reflective thought, setting rules for self-control) (Johnson et al., 2012). This characterization of nudges does not fit well with what practitioners who are involved in informed decision making do (i.e., we aren't simply nudging people—we are informing people so they understand and can autonomously make their own decision). We take nudges to target a much narrower band of interventions aimed at improving behavior through non-rational persuasion.

Paternalistic policies influence people's decisions while leaving alternative choices genuinely open without penalty or other incentives. Nudges that are Libertarian Paternalistic are paternalistic in that the nudges try to alter choices. Those nudges are libertarian in the sense that alternative choices are available without changes to incentives. Ultimately, however, Libertarian Paternalism is a type of (soft) paternalistic policy that involves some moral violation. So, ethically speaking, these kinds of nudges require the same kinds of justification that would be required for any other type of paternalistic policy.

It may appear that Libertarian Paternalism is identical with Choice Architecture. They both change the decisions that people make in predictable directions. But as we understand the two concepts, there is an important distinction. Recall that paternalism involves a moral violation, such as violations of autonomy. Yet instances of Choice Architecture do not necessarily involve any moral violation. To illustrate, providing people with accurate and relevant information that predictably changes decisions qualifies as a type of Choice Architecture. But informing people's decision need not involve a moral violation because that information could be integrated into rational agency and actually promote autonomy (Johnson et al., 2012; Muenscher et al., 2016). In those instances, providing information does not violate a moral rule and hence is not a kind of paternalism (much less Libertarian Paternalism). However, other types of Choice Architecture can be justifiably characterized as Libertarian Paternalism in cases when interventions bypass rational agency (e.g., nudging people toward a choice by taking advantage of automatic processing alone, as happens when people influence choices by setting opt-in or opt-out defaults). Thus, Choice Architecture can, but does not necessarily, involve a moral violation and so should not be viewed as synonymous with Libertarian Paternalism. An illustrative model of the potential relations between Choice Architecture, Nudges, and Libertarian Paternalism is displayed in Fig. 7.1.

On this model, Choice Architecture is the broadest kind of way to alter people's choices (e.g., any behavioral interactions in the world can be described in terms of its Choice Architecture). Choice Architecture can be characterized as the environment in which people make decisions that may influence choices. Many instances of Choice Architecture are naturally occurring and non-intentional (e.g., sunlight can influence moods and related behaviors). A more specific, proper subset of Choice Architecture is nudging. Among the distinctions between nudging and choice

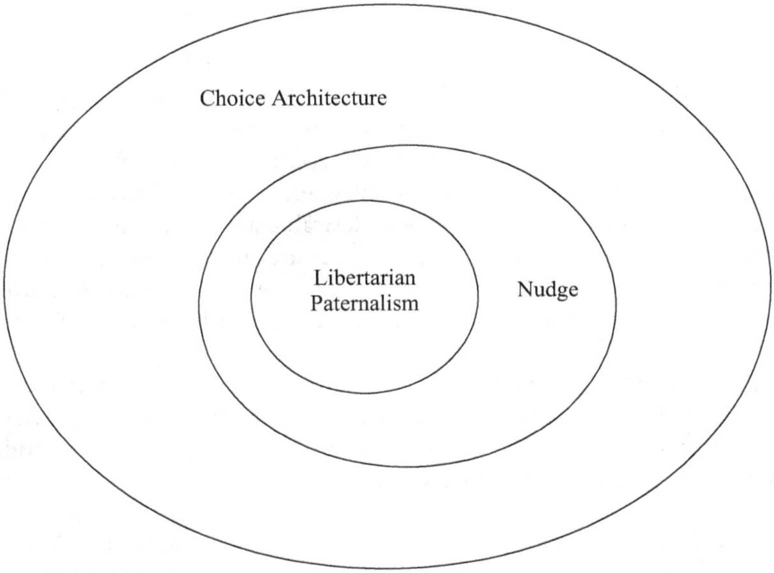

Fig. 7.1 Conceptual diagram of distinctions and nested relations among choice architecture, nudging, and libertarian paternalism

architecture is that nudging in some way intentionally structures the decision making environment to promote a specific choice. For example, a dried riverbed (i.e., a natural environment) can change the path you walk just like a ditch could. But the ditch could be intentionally placed to alter your choice, nudging you down a different path. Finally, an even more specific kind of choice architecture is Libertarian Paternalism, a proper subset of Nudges (and, by transitivity, a proper subset of Choice Architecture). Libertarian Paternalistic policies involve some (perhaps justifiable) moral violation, whereas nudges need not (e.g., some nudges are transparent and engage rational agency). While there is still no wide consensus about these conceptual distinctions, our working assumption is that the most controversial kinds of choice architecture are those that are Libertarian Paternalistic because they involve some moral violation. These kinds of choice architectures (and nudges) will be the focus of the rest of this chapter unless specifically noted.

The Ethics of Libertarian Paternalism

The ethics of Libertarian Paternalism are hotly debated (Blumenthal-Barby & Burroughs, 2012; Hausman & Welch, 2010; Welch, 2013). Almost everyone agrees that in some circumstances, paternalistic policies that influence choices are justified and sometimes even necessary. For example, it is relatively morally uncontroversial that, on average, having more organs available for transplant is desirable and having fewer deaths is preferable to having more deaths. And, the argument goes, almost everybody values those outcomes. So, the ethical costs associated with nudging people to be organ donors are justifiable given sufficiently good enough outcomes.

However, even in the instances where the good seems to outweigh the cost of the libertarian paternalistic policy, there is still a cost. In particular, one cost that is commonly identified with libertarian paternalistic policies is that those policies undermine autonomy. To illustrate,

> To the extent that they [nudges] are attempts to undermine the individual's control over her own deliberation, as well as her ability to assess for herself alternatives, they are prima facie as threatening to liberty, broadly understood, as is overt coercion. (Hausman & Welch, 2010, p. 131)

The worry is that through the predictable influence of non-rational features, one's choice can be influenced by the intentions of another person via the nudge: "Their actions reflect the tactics of the choice architect rather than exclusively their own evaluation of alternatives" (Hausman & Welch, 2010, p. 128). As such, the nudge could be coercive. In such instances, the first condition of Mele's sufficient conditions for autonomy is not satisfied. The nudge could also run afoul on Mele's second condition where the person has all the relevant information to make a decision. The choice that is nudged is a function of the information that is strategically provided. For many nudges, there is no intent or effort to ensure that the person would have a minimally sufficient set of the relevant information to make an informed decision (i.e., a representative understanding). Rather, the information is an intentionally small (e.g., skewed or biased) subset of the relevant information. In such instances of nudging, that small set of biased information increases the probability that people will make the "desired" decision. Hence, one potential path to autonomy is not secured, and a moral rule is violated in the Libertarian Paternalistic policies.

Providing appropriate justification for Libertarian Paternalism is complicated, especially when there are different values at stake. If what we have presented throughout this book is correct, then values will often be diverse and stably related to one's personality. The diversity generates two challenges for libertarian paternalistic policies. One challenge is internal to Libertarian Paternalism:

> *Internal Challenge:* Sometimes, it is not clear what values we should promote given the diversity of values (Hansen & Jespersen, 2013; N. Smith, Goldstein, & Johnson, 2013).

Value diversity presents a common problem for policies that attempt to predictably alter decisions in some direction because it is often contentious what that direction should be.

Our data also present an external challenge to Libertarian Paternalism:

> *External Challenge:* Libertarian Paternalism is not simply justified if the (direct) goods that are generated by the policy are outweighed by the (direct) moral costs of the policy.

The External Challenge requires a bit of elucidation. One might think that if the benefit of the Libertarian Paternalistic policy is higher than the moral costs, then one should institute the Libertarian Paternalistic policy. However, looking only at the costs and benefits of the Libertarian Paternalistic policy leaves out an important element in policy decisions. Namely, whether there are other policies that could be instituted that generate similar benefits but without similar costs. If one only focuses on the good consequences and the opportunity to nudge, one may miss the opportunity to evaluate the relative benefits as compared to other potentially powerful alternatives (and relative base-rates and mechanisms of each—i.e., including why, when, and from whom the various policies succeed). Ultimately, if there were other strategies that could achieve the same, or similar, ends as the paternalistic strategy but that did not violate a moral rule, then there would be little ethical or practical justification to opt for the paternalistic policy. Thus, ethically it follows that the relative costs/benefits of Libertarian Paternalism should be compared to alternative strategies that could achieve the same or similar ends. To more fully map these issues, we next consider one potential alternative to Libertarian Paternalistic policies.

Ethical Interaction Theory: Interactive Policy Analysis

Ethical Interaction Theory is a framework designed to offer techniques to quantify and compare ethical costs and risks associated with individual instances of choice architecture. In this light, our preferred alternative to Libertarian Paternalistic policies is to inform people, empowering them to make decisions on their own—a kind if informing we call *representative education* (cf. "Boosting," Hertwig & Grüne-Yanoff, 2017). We will discuss more what we take *representative education* to be below. But the main thrust of the approach is to provide people with enough relevant information so that they have a high-quality factual base to make decisions (e.g., promoting a representative understanding, see Cokely et al., 2018). Informing people is nothing new. However, there has been considerable debate about determining what constitutes relevant quantity and quality of information (Berleur, Nurminen, & Impagliazzo, 2006; DiPazz, 2002; Turilli & Floridi, 2009; Winkler, 2000). We propose that representative education offers a new solution to the quantity and quality problem.

Let's consider issues with quantity of relevant information first. Complete information could theoretically avoid problems associated with the intentional use of non-rational factors to influence decisions. If one knows everything relevant to the decision and if one could integrate all of that information into a decision, then there would be no need for non-rational factors to play a role in the decision. Even if non-rational factors could potentially play a role in decision making, the information about those non-rational factors would simply be one more informational input in the decision. Those non-rational factors would no longer be non-rational factors since they would be integrated in the decision making process. Of course, a major worry is that if the necessary quantity of information is sufficiently high, autonomous decision making might rarely be an achievable standard for humans (Gigerenzer et al., 1999; Hardman & Macchi, 2003; Merz & Fischhoff, 1990). Think about the last time you sought a mortgage for your home or the last time you consented to a medical treatment or even surfed for a TV program. Odds are you had access to *a lot* of relevant information that you probably didn't or perhaps could not explore. Clearly, complete information is often rather impractical if not impossible.

If providing all information about a decision is neither a necessary condition nor part of a sufficient condition for autonomy, then it appears that the *quality* of the information is what matters most. We think that providing information that a person can efficiently integrate into a *representative understanding* of the decision problem is key to autonomous, adaptive, and informed decision making. In particular, if we can present information in ways that facilitate representative understanding, then we will promote autonomy compared to instances where people developed a biased or unbalanced understanding based on systematically skewed, persuasive information. That increase in the representative quality of the understanding may then contribute to that person's autonomy, not because of full information but because one has a more prognostic understanding of the decision problem and how it factors into one's own life and values. Accordingly, developing valid and robust scientific means and methods for assessing, characterizing, and evaluating such representations (e.g., costs/benefits, robustness, trade-offs) is a central enterprise in the science for informed decision making and a major marker of scientific maturity and progress (e.g., increasing prediction and control of representative understanding and decision vulnerabilities).

Generally, a representative understanding happens when one's understanding is relatively robust against bias given random additional relevant or irrelevant aspects of information. Bias, in the sense that we mean here, only implies a tendency, not an error (e.g., many Americans have a bias to write with their right hand). More specifically, representative understanding (a person variable) as well as representative education (an interface variable) can be understood by an analogy to representative sampling for statistical inference. A sample is representative of a population when the sample accurately reflects the target population on the properties of interest (e.g., by having sufficiently large, random sample). When the sample is representative (and not too small), adding additional randomly selected data will not likely change the robustness of inferences made about the population, assuming appropriate statistical techniques are used and standard assumptions are met. Likewise, having a representative understanding means that one's understanding is sufficiently (but not exhaustively) nuanced and detailed such that additional random aspects of information are unlikely to bias inferences made on the bases of that understanding. Following the sampling analogy, any random bit of information (either relevant or not, accurate or erroneous) is not likely to change one's mind if one has a representative understanding. To the extent that one's decisions are easily (or substantially) biased by additional random information, then one does not have a representative understanding. To the extent that more representative education changes the decision, one does not have a representative understanding.

To offer an oversimplified but potentially useful illustration, consider making a decision about getting burned. Once a person realizes that a flame burns and hurts their finger, most adults have sufficient personal understanding, knowledge, and reasoning capacities to develop a relatively representative understanding of key causal aspects of the relationship between fire and the rest of their body surfaces (e.g., fire burns and hurts my finger; even though some skin areas are less sensitive than fingers, something that hurts the skin of my finger will likely hurt the skin in another part of my body. Thus, fire will probably hurt anywhere it touches my skin). With this understanding they can make reasonable, robust inferences about how much they (don't) want a flame to touch them elsewhere. By chance or intent, they may come across more information that could cause them to update their previous understanding. But, absent a relatively concerted and compelling effort to manipulate incoming information or to discount one's previous knowledge about one's self or one's environment (i.e., accurate Bayesian priors), the simplified yet representative causal understanding of skin and fire (a representative sample) will tend to allow them to use simple decision strategies (heuristics) to make inferences that approximate the decisions they would make if they had an expert understanding of all the relevant information (the population).

Because representative understanding essentially involves rational agency, ensuring representative understanding is autonomy promoting. For this reason, any strategy that informs, even if slightly, is to be preferred to a strategy that does not inform, everything else being equal. And, the informing is autonomy promoting because people are free to integrate that information with whatever (diverse yet stable) values they may have. In this way, promoting representative understanding through education can be different from nudging. Nudges focus on the *outcome* of a decision process (e.g., eating a salad, installing energy-efficient light bulbs), whereas promoting representative understanding through education focuses on the *decision making process* (e.g., rational agency) that leads to the outcome.

In the light of theory and extant data, it is clearly *possible* to avoid many problematic aspects of and debates about Libertarian Paternalism by providing representative education. But we want to be very clear that *the possibility does not entail that we should always prefer promoting representative understanding to Libertarian Paternalism.* By our lights, the representative education framework does not imply that any specific class of decision policy is always best. Rather, the framework offers a conceptual basis for an ethically and empirically informed interactive policy analysis of the relative merits of viable options (e.g., comparing best practice Libertarian Paternalism to decision aids or educational interventions). As the science and practice matures, we expect this process will ultimately follow standards such as formal costs-benefit or policy analysis (e.g., Gramlich, 1990; Weimer & Vinning, 2017).

To further briefly clarify, let's consider what such a cost-benefit analysis would look like and what dimensions would be relevant. While a complete account is still to be discovered and established, we can give the contours of what an analysis would look like. An efficient starting point is Trout's (2005) suggestion concerning strategies that attempt to debias and inform (i.e., internal strategies) as compared to those that influence by taking advantage of the biases (i.e., external strategies):

> To the extent that these particular strategies work, their desirability is based on the particular features of the problem: their generality (the scope of the problems they address), their frequency (how frequently the types of problems they address actually occur), their significance (how important the problems are to human welfare), and the cost of implementation (how simply and cheaply the problem can be addressed by these methods). (Trout, 2005, p. 422)[5]

In some instances, promoting representative understanding will be superior to Libertarian Paternalism on these criteria. Take the implementation criterion first. There are some simple, efficient, and direct ways to increase representative understanding. The presentation of visual aids has been shown to increase understanding of basic information relevant to some decisions (Galesic & Garcia-Retamero, 2011; Garcia-Retamero, Petrova, Feltz, & Cokely, 2017). This kind of intervention is arguably

[5] Similar considerations for choosing a debiasing strategy are offered by Soll, Milkman, and Payne (2015). They recommend (1) Evaluating the relative effectiveness of the debiasing strategies, (2) Determining the decision readiness of the individual (i.e., are they tired, do they have the relevant skills, etc.), (3) Assess heterogeneity of values, (4) Estimate the decision frequency, and (5) Estimate the decision complexity.

often at least as easy to implement as structuring environments so that the message frame or defaults influence people in the desired direction.

Second, interventions that successfully inform decisions (e.g., promote representative understanding) may confer other general benefits that are missed by the narrowly focused Libertarians Paternalistic strategies. Informing people's decisions engages rational agency in ways that Libertarian Paternalistic policies do not and therefore has the potential to encourage the development of character, skills, and wisdom that may be valuable or even essential for generally skilled decision making, personal growth, and more comprehensive rational agency. For example, perhaps bar graphs provide the opportunity for people to understand statistical information better, and that experience and familiarity with bar graphs may transfer to other, similar decisions that involve statistical information. Alternatively, perhaps careful evaluation of trade-offs and options, e.g., in high-stakes medical contexts, provides people with greater insight into their own deeply held values, or a greater sense of decision making self-efficacy. In any event, because Libertarian Paternalistic policies do not increase understanding or agency to the same extent as efforts to increase representative understanding (if at all), Libertarian Paternalistic policies are not likely to help nurture these powerful kinds of skills, insights, or resources, and thereby do not generally promote autonomy or rational agency distally or proximally.

Third, Libertarian Paternalistic policies derive their effects from interventions targeting relatively passive decision making, and thus the quality of decision outcomes depends on the wisdom and power of policy makers to shape environments in suitable ways. That is, Libertarian Paternalistic policies take time and resources, including political capital, to implement. Libertarian Paternalistic policies are also typically only effective under a narrow band of conditions (e.g., under routine conditions when everyone has similar biases and would benefit from similar outcomes). What's more, Libertarian Paternalistic policies appear to run a significant risk of encouraging more passive, dependent decision making more generally (e.g., passive behavior is reinforced and rewarded). Even if the risk is small in any single instance, given enough time and exposure, Libertarian Paternalistic policies appear likely to reduce one's decision making self-efficacy, potentially damaging one's deep sense of competency. Factors that threaten self-efficacy and agency in turn tend to undermine motivation toward and resiliency of autonomous behavior, innovation, well-being, creativity, leadership, skill development, and a host of other factors with real social and economic implications. In contrast, efforts to develop more autonomous decision makers theoretically should promote personal development and agency, promoting more adaptive, self-determined, skilled, and resilient decision making more generally. In such cases, autonomous and skilled decision makers are individuals who are well-equipped to make good decisions for themselves, their families, and their communities. Those kinds of decision making skills and abilities are likely to provide even larger benefits when decision makers face rapidly changing, high-stakes, and evolving conditions (e.g., when infrastructure becomes less reliable or when threats escalate, such as during natural disasters, emergencies, and in unfamiliar social and economic conditions). Underlying skills and abilities also appear to provide some protection against the threat of misinformation and disinformation and may also reduce people's susceptibility to motivated reasoning biases that often follow from conflicts of interest, particularly in controversial domains (e.g., Climate Change; for a recent review see Cho et al., 2024; see also Van der Linden, 2023; Roozenbeek, Van der Linden, 2024).

Of course, efforts to increase representative understanding are not free of problems. Representative education can be expensive in many senses (e.g., time, money, and other resources) and may not be as effective or efficient as Libertarian Paternalistic policies, particularly in some high-stakes instances (see Feltz (2015a, 2015b) and Trout (2005)). Nevertheless, the extant data consistently indicates that systems that promote skilled and informed decision making tend to empower autonomous, high-quality decision making. Given the overwhelming evidence on the mechanisms and value of skilled decision making and related outcomes (e.g., resiliency, well-being, and agency), even if the short-term costs and benefits are relatively comparable, we can be confident that autonomous decision making will usually be morally and practically preferable to Libertarian Paternalistic policies.

To illustrate, Benartzi et al. (2017) have conducted a comparison of different types of choice architecture including Libertarian Paternalistic policies and more hard paternalistic strategies like offering monetary incentives for some choices. For instance, they measured the effectiveness of a default nudge for flu vaccination versus monetary compensation for taking part in a flu vaccine. They found that about twice as many adults got a flu vaccine in the default condition compared to the monetary incentive condition (per \$100 spent). Hence, it looks like on the surface the default nudge is more effective than monetary incentives. However, we re-analyzed their data to respect the distinctions between choice architecture, nudges, and libertarian paternalistic strategies (the authors lumped all choice architectures besides incentives into one group). On our re-analysis, strategies that involved informing individuals (e.g., educational campaigns) were three times better than libertarian paternalistic nudges and eight times better than monetary incentives. This same pattern held true not only for decisions about the flu vaccine, but also for decisions about saving energy and retirement savings. Hence, in these cases, it appears that the benefits of informing greatly outweigh the benefits of the libertarian paternalistic policies independent of promoting autonomy. Critically, we would not have known how much better these policies were had we not compared them.

Choice Architecture Policy Analysis: A Detailed Case Study

To illustrate more concretely key aspects of an interactive policy analysis, we consider one case study comparing choice architectures used for risk communications (e.g., libertarian paternalism v. information transparency). We aim to illustrate some side-by-side comparisons of different choice architectures that are instructive with respect to how, when, and why we can infer that libertarian paternalistic decision policies are ethically inferior to informed decision policies (e.g., representative education that promotes representative understanding).

Our example comes from a line of research that attempts to promote better decision making related to sexual health and disease prevention. As we have already discussed, the way that choices are framed (i.e., how information is described) can predictably influence the choices that some people make, even if the information presented is formally logically identical (Levin, Johnson, & Davis, 1987; Mcneil, Pauker, Sox, & Tversky, 1982; Rothman & Salovey, 1997; Tversky & Kahneman, 1981). In a series of experiments, Garcia-Retamero and Cokely (2015a, 2015b) demonstrated that sexually transmitted infections (STI), relevant choices, and behaviors were predictably altered depending on how risk communications framed the risk information. In one part of the study, participants were randomly assigned to receive "positively" or "negatively" framed information about the risks associated with STIs. The "positive frame" emphasized that using condoms *reduced* the chances of contracting an illness and having long-term health consequences. The "negative frame" described the same basic information in terms of *increasing* the chances of contracting illnesses and long-term illness if one *does not* use condoms. Remarkably, this small change in description had large impacts on the resulting screening or protective behaviors of the young adult sample involved in the study (note: young adults are among the most at risk for life-altering HIV and related STI infections that could be largely prevented with condom use). Those who received the information positively framed were nudged toward engaging in more preventative behaviors (e.g., using a condom) than those who received the negatively framed information. Those who were given the information negatively framed were more likely to engage in screening behavior (e.g., getting a test for an STI) than those given the positively framed information.

Given these results, the science indicates that a choice architecture policy using gain and loss framing for information about STI prevention and decision making can successful, quickly, and robustly decrease some risky behavior (e.g., promote condom use or STI screening). Theoretically, this type of choice architecture represents a libertarian paternalistic policy as the framing of the information is designed to encourage people to engage in the targeted behavior while preserving the ability to choose. Thus, even in the absence of representative education (and the resulting representative understanding) these nudges appear to have some beneficial effects. Taken at face value, these results may suggest framing information to maximize one of those outcomes (prevention or screening) may be ethically defensible. How do these effects compare to other choice architectures using representative education to promote more representative understanding about detection and prevention costs and benefits?

To estimate the relative benefits of information transparency, Garcia-Retamero and Cokely (2015a, 2015b) gave participants data visualization as a decision aid (see Fig. 7.2). The visual aid took the form of a simple bar graph that depicted the statistical information about risk and reduction of infection in a visually accessible (i.e., transparent) and easily comparable form. All other information was the same (e.g., presented with negatively or positively framed information), yet giving participants this visual aid resulted in no measurable differences in screening and prevention behaviors as a function of message framing. Even though there was no measurable difference between the gain and loss framing on screening and prevention behaviors, the pair of behaviors together were dramatically influenced when the graph was presented compared to when the graphs

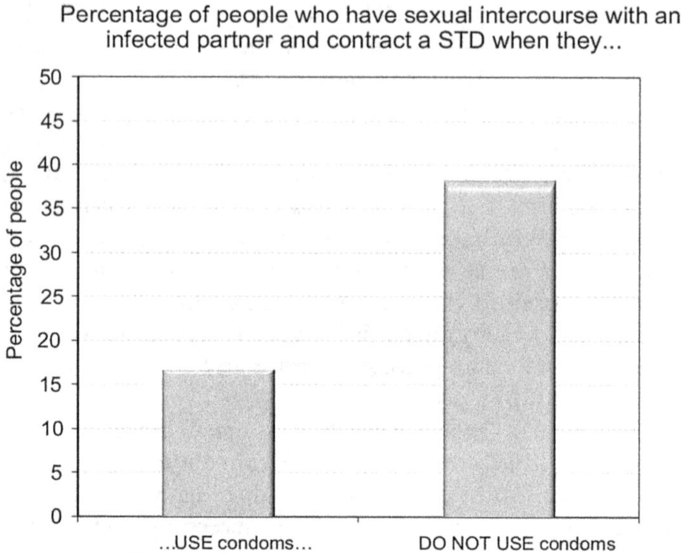

Fig. 7.2 Visual representation of infection rates with and without using condoms

were not provided (e.g., reductions of risks in subsequent behaviors). Thus, visual aids encouraged people to engage in increased amounts of prevention and screening behaviors regardless of the frame when provided with the transparent visual aid bar graph (i.e., just as highly effective as either framing intervention, on average). On the face of it, it is reasonable to assume that those who were provided with the visual aid generally understood the decision problem better than those who did not receive the decision aid. Hence, on our view, the aided decision is ethically better than the nudged decision.

One could think that the presentation of the visual aid is simply one more libertarian paternalistic method to nudge people to the desired choices. That is, perhaps there is something about the graph that takes advantage of non-rational features in order to increase the desired screening and prevention behavior. But there are some important and telling clues suggesting that this is not how the graph worked. Garcia-Retamero and Cokely (2015a, 2015b) assessed how well participants understood the information about sexually transmitted diseases. When people were given the visual aid, they understood and could reason about the information better than when they were given only the framed information, which translated into enduring changes in attitudes, plans (e.g., intending to buy condoms or visit a physician), and behaviors. This is consistent with a large literature on the effect that visual aids can have on improving information understanding (Garcia-Retamero & Cokely, 2013a, 2013b, 2013c; Garcia-Retamero, Cokely, & Hoffrage, 2015; Garcia-Retamero, Okan, & Cokely, 2012). Moreover, in subsequent study, a similar design was used to test the benefits of visual aids compared to other kinds of framing effects (e.g., attribute instead of gain/loss framing) or compared to a validated and extensive (eight hour) educational intervention (Garcia-Retamero & Cokely, 2013a, 2013b, 2013c, 2017). In both cases, the simple visual aid generally matched the large benefits of other framing-based choice architectures, *and* of the extensive educational intervention, requiring equal or lesser amounts of time and costs to implement, and resulting in similarly enduring changes in target attitudes, intentions, and behaviors. Importantly, however, the results also revealed that the visual aids improved representative understanding to the greatest extent among most vulnerable individuals (e.g., individuals with less knowledge of risks and lower risk literacy scores as measured by numeracy tests)—largely eliminating disparities by roughly equating more and less skilled decision makers on most assessed decision making quality variables. Ultimately, representative understanding of information is a rational factor that tends to be by far the most influential variable that gives rise to better decision making. So, at least with respect to framing information, providing visual aids may, and certainly does in some cases, provide a more representative understanding of the problem than the libertarian paternalistic policy.

Given the results of this randomized control trial and the associated estimates of the comparative value of the two architectures, we can directly compare some key costs, benefits, and potential trade-offs of the two decision policies. The visual aid was ethically superior because it protected (rather than infringed on) autonomy. Other things equal, in accord with the Ethical Interaction framework, this alone would imply that the representative education policies should be preferred. But, of course, other things aren't often equal and so we next turn to other costs and benefits that would be relevant for wider-scale implementation of both policies.

Roughly, the production and implementation costs and benefits of both the visual aids and framed information appear likely to be similar (e.g., printing, posting, and distribution), although the inclusion of a basic bar graph simplifies brochure development in the decision aid condition (e.g., doesn't require the use of two separate brochures using gain frames for prevention and then loss frame for detection). The statistical model estimates from the experimental trial indicate a relative equivalence across the direct effects on decision outcomes and targeted consequences. However, the *cognitive and emotional* costs and benefits, such as the attention and processing time (i.e., reading) and overall cognitive workload, ironically may be lower for the visual aid condition, which is theoretically easier to use and remember (e.g., a picture is worth a thousand words). Moreover, the visual aid did not explicitly aim to induce mood-type states that may entail other non-rational carry-over effects that could be hard to counteract without a more representative understanding (e.g., inducing risk or loss aversion more generally). To the extent that the visual aids are likely to be easier to communicate and remember, they should also be at least marginally easier to accurately discuss with others and more resistant to distortion effects (e.g., misremembering).

Interestingly, people who were less numerate and less knowledgeable appeared to be affected by both the framing and visual aid manipulations (see Garcia-Retamero and Cokely (2013a, 2013b, 2013c)). Because less numerate people are also less prepared to independently evaluate information about risk (e.g., they have lower risk literacy), the framing manipulation also appears likely to create some disparity in the cognitive and decision making benefits in libertarian paternalistic (framing) conditions. Essentially, even in the framing condition, more numerate people are likely to use their risk literacy skills (e.g., reframing) to generate a more representative understanding of the underlying information. This processing makes it more likely that those who are risk literate will avoid being affected by framing but those with lower risk literacy won't. Hence, there are disparities with respect to whose autonomy is diminished. To the extent framing effects do not influence decision making for numerate people but do for less numerate people, we have yet another ethical concern that was circumvented in the visual aid condition (e.g., all individuals who made better decisions were more likely to do so on the basis of a more representative understanding—roughly equating risk literacy for the current decision).

Taken together, the net benefits of the representative education policy (i.e., transparent visual aids) seem to dominate those of the libertarian paternalistic policy (i.e., the framing manipulation), without any appreciable trade-offs. Most implementation costs were similar, yet the decision aid protected autonomy and provided a more enduring means of empowerment that was shared more equitably among more and less vulnerable individuals (e.g., promoting informed and skilled decision making across all, instead of biasing less skilled individuals while informing others). In accord with the standards of Ethical Interaction Theory, in this case the representative education policy is both ethically and practically superior.

As this example suggests, in some instances there are important and viable alternatives to nudging. To put the point somewhat differently, at one time in the recent past the available science suggested "there is no evidence that the same corrective success could be effectively or routinely achieved by inside strategies, strategies of individual motivation that attempt to acquire more accurate representations or to consider alternative possibilities" (J.D. Trout, 2005, p. 430). Since then, decision aids and training programs have been shown to hold great promise with regard to the promotion of informed decision making. And these educational interventions did not "cost" more than nudges. Thus, we do not need to consistently or exclusively rely on nudges to help people make better decisions. That said, technologies that promote representative understanding also do not always result in better decisions in instances where nudges are effective, and in some cases the costs associated with informed decision making may far outweigh any benefits. The task that is left to us, then, is to determine when nudges are ethically preferable to alternative strategies like representative education. We provide some preliminary criteria to use in comparisons in the next section.

Choice Architecture Policy Analysis: Heuristic Evaluation

We have reviewed theory on some of the relative merits of nudging versus promoting representative understanding. In efforts to distill some practically useful and efficient guidelines, we next consider five primary concerns for heuristic evaluation by choice architects, designers, and policy makers. Each of these heuristics may be useful when trying to estimate the relative costs and benefits of various choice architectures. Based on the previous discussion, there is reason to think this heuristic evaluation will be a robust and practical (but not necessarily perfect) guide for complex interactive policy analysis.[6] But, to be clear, this is only a starting place (see Appendix), as there just is not enough science on nudges or their alternatives to propose or justify a full set of formal standards. Given these considerations, the following is a tentative list of heuristics to use when comparing nudges or informing decision makers. These heuristics are Benchmarks, Disparities, Resources, Reputation, and Resiliency.

1. *Benchmarks: What are the alternatives and benchmarks against which we should evaluate the benefits of nudges or other decision policies, and why?* Whenever there are potential or actual alternatives to libertarian paternalism, then an interactive policy evaluation of the relative costs and benefits of the libertarian policy should be conducted. We have already discussed some of relent criteria (e.g., those provided Trout (2005)). One additional potential evaluation element should involve an assessment of the moral costs associated with alternatives. Recall that one of the central characteristics of paternalistic policies

[6] When appropriate, feasible, and robust, designers and interactive policy analysts may want to consider well-established formal methods (e.g., cost-benefit analysis, policy and econometric analysis, applied decision analysis; see Boardman, Greenberg, Vining, & Weimer, 2011). However, because the ethical decision support approach is relatively new and includes many underspecified elements, caution is merited when attempting to precisely integrate and evaluate various aspects of different policies. Expert and formal efforts notwithstanding, the complexity and uncertainty involved in any such endeavor carries well-understood, yet difficult to manage risk (e.g., overfitting and biased estimation). For these and other practical reasons, wherever there are large differences in outcomes or costs, heuristic evaluation methods are likely to be useful and effective. Heuristic-based evaluations (e.g., expert evaluations, checklists) have a long history of useful application in design and engineering fields, including human factors, human-computer interaction, and consumer product development domains. In short, these are methods that direct the evaluator to note and characterize key attributes or issues. These simple evaluation methods are often useful even when used with so-called naïve simple heuristics like unweighted tallying as in Franklyn's rule (e.g., listing unweighted pros and cons and then selecting the option with the higher ratio of pros to cons). Of course, it may sometimes be reasonable to weight (value) the various attributes (pros/cons) and consider the estimated magnitude of trade-offs, or even integrate using some function as in a multi-attribute utility analysis (e.g., a scaled-up heuristic evaluation that approximates more formal decision analytic methods). Ironically, however, in many naturalistic environments where there is low stability, high complexity, and/or uncertainty (i.e., the "real world"), increasing the precision of the analysis may decrease the predictive accuracy, thus undermining the utility of the analysis (e.g., bias-variance dilemma in statistics and machine learning). Absent good justification and/or expertise, ethically informed simple heuristic methods for decision policy analysis seem likely to provide useful, efficient, and reliable insights (see also Gigerenzer et al. 1999; Gigerenzer & Brighton, 2009).

is that they involve some moral violation, which is not an issue for some alternative policies (e.g., transparent decision aids that promote representative understanding). Therefore, in order to ethically justify using a libertarian paternalistic policy, the libertarian paternalistic policy must perform better on relevant criteria compared to the alternative in the light of other relevant associated costs and pragmatic concerns (e.g., given costs, needs, constraints, etc.). Detailing these moral and other costs and benefits is necessary (but not sufficient) for the establishment of ethically defensible nudging policies.

2. *Disparities: How widely do stakeholders and experts agree about the value(s) being promoted? Are common differences in fundamental values likely to result in conflicts of interest?* Libertarian Paternalistic policies typically find their most persuasive support in instances where there are uncontroversial normatively correct choices. The main reason why is that Libertarian Paternalistic policies, by their very nature, promote only one choice.[7] For example, default settings only attempt to encourage the default congruent choice. If there is only one basic value, then Libertarian Paternalistic policies can sometimes efficiently help secure that value. However, as we have argued, values are often diverse, stably related to personality, and there is not only one value that should be maximized above other values (e.g., different people with different end-of-life values may have different priorities when designating a surrogate). Given that there typically is no known Neo-Platonic truth about what values are correct (or even which most are most valuable), choice architects won't be able to reliably guide people to make decisions in

[7] Some have argued that we should deploy *smart defaults* (N. Smith et al., 2013). Smart defaults build a profile of individuals to make tailored predictions about what defaults would be most appropriate for those individuals to maximize good decision making for that individual in light of the values that individual might have. In some instances, Libertarian Paternalistic policies could be set only for those who would benefit from them. In this way, smart defaults avoid the one-size-fits-all criticism. This is an interesting notion, but there are the same fundamental problems associated with the smart defaults being paternalistic (if the smart nudge is indeed a paternalistic strategy). Moreover, smart defaults are effort and resource intensive because the profiles often take quite a bit of effort to become fine-grained predictors—as indicated when N. Smith et al. (2013) write that these kinds of nudges would have to "understand consumers better than they understand themselves" (p. 167). As such, smart defaults run the risk of reducing the advantage that nudges have with respect to efficiency of implementation.

accord with the "right" values whenever there is diversity of legitimate values. In cases such as these, Dworkin (1988) gives some helpful suggestions about how to weigh the alternatives to nudging when there is a plurality of values:

1. The majority interest must be important
2. The imposition on the minority must be relatively minor
3. The administrative and economic cost of not imposing on the minority would be very high.

To the extent that the there is good evidence that 1–3 are true, then that may generally favor some Libertarian Paternalistic policies (Soll et al., 2015). However, in instances where there is little or weak evidence for some of 1–3, then one must think carefully about the relative merits of the Libertarian Paternalistic policy versus alternative choices.

3. *Resources: What are the essential skills and resources required for the individual decision maker, how well understood are those competencies, and what are the distributions of relevant competencies across various stakeholders?* In general, Libertarian Paternalistic strategies will have the advantage when we are not sure what skills are required in order to make an independent, well-informed decision or when developing those skills is prohibitively costly (e.g., becoming an expert violinist). There is a substantial literature that suggests there are at least some domain-general skills that lead to better decision making in general (e.g., statistical numeracy is thought to give rise to risk literacy more generally; Cokely et al., 2012), as well as many other domain-specific competencies that can powerfully influence choice (e.g., expertise, time, financial resources, familiarity with the domain, etc.). Choice architects who have identified the relevant skills and resources (e.g., reverse engineering superior decision making) can more accurately assess the feasibility and other costs associated with various designs (e.g., by "boosting" competencies (Hertwig & Grune-Yanoff, 2017)). For example, it would be ethically irresponsible to design a decision education intervention that no one would understand or have time to consider (e.g., too complex), given the availability of effective and otherwise useful nudges.
4. *Reputation: What are the costs to the Choice Architect?* All the other criteria in this heuristic evaluation list concern the end-user who is the target of the instance of Choice Architecture. However, atten-

tion should also be paid to the costs to the Choice Architect who is implementing the decision making intervention. Sometimes, choosing to intervene comes with risk to the intervener. These risks could take a variety of forms including reputation costs, trust (e.g., "trust in municipal authorities"), and, in some cases, risks of physical harms (e.g., interventions to decrease homophobia) (2011). There is little research on the effects of Libertarian Paternalistic policies with respect to what we call reputation costs and how those nudges influence those factors. Some research exists about how general factors like one's level of trust predicts acceptance of nudges, suggesting that greater institutional trust predicts acceptance of nudges (Sunstein, Reisch, & Kaiser, 2019). Other research suggests that nudges that target unconscious processes are viewed as less favorable than those that target conscious processing (Felsen et al., 2013). However, little research exists indicating whether nudging influences levels of trust in the choice architect or their agency (Hoang & Feltz, in prep).

5. *Resiliency: What are the relevant infrastructure, development, and similar constraints that bear on the feasibility of each instance of choice architecture?* Special attention should be paid to the practical, legal, and implementation costs associated with deploying any choice architecture with respect to potentially dramatic changes in the choice environment (e.g., including cognitive or emotional changes within the decision maker or social-political changes in the environment). This is particularly true for high-stakes and time-sensitive choice architectures that rely on special kinds of infrastructure or resources (e.g., changing administrations can change priorities and reduce access to resources; cyberattacks can alter communication infrastructure and cognitive workload; natural disasters can alter physical infrastructure and emotional stability). For instance, it may be wise to use a Libertarian Paternalistic policy to help people make better choices rather than to explore options about how to inform in some deadly situations when public risk perceptions could interact to exacerbate the emergency (e.g., when fear can dramatically alter behavior). If there is an outbreak of some disease that is highly contagious, perhaps it is better to set up defaults (or even hard paternalistic policies) that keep people out of harm's way. In those instances, the gains made in terms of expediency could outweigh the costs of the moral violation even if there are, at least in principle,

other alternatives that theoretically could inform decisions (e.g., risk communications that should be transparent but cannot be validated in time). Of course, in similar emergency situations that may involve rapidly changing or persistently unstable conditions, decision aids that enable independent informed decision making and non-technology-mediated person-to-person communication may be favorable, as might be the case after a natural disaster that involves long-term utility and transportation disruptions. In any case, choice architectures should assess all relevant aspects of infrastructure needs and vulnerabilities, with careful attention to risks that accrue under changing cognitive, emotional, social, political, and environmental conditions.

Conclusions

This book had an ambitious goal: to provide an integrative evidence-based review of the growing literature showing that many philosophical values are predictably fragmented while also detailing how and why that fragmentation has some important theoretical and practical implications. We started by documenting the varied yet predictable nature of many people's basic philosophical and ethical values. The main philosophical conclusion drawn from all relevant studies was that some philosophical projects run the serious risk of not being able to be reliably done based on the current methods and tools (e.g., Neo-Platonic projects). In the light of other research, it follows that it is unlikely that we will ever come to know with any great confidence the mind-independent truth about a variety of philosophical issues including essential truths about freedom, moral rightness, intentional action, and some of the values we hold most dear.

Given this analysis, what we are left with is a variety of different concepts and values that all seem, at least on the face of it, to be acceptable and normal for people to have. In the presence of irreducible and justifiable diversity and fragmentation, what are we to do? In closing our book, we have focused on providing a theoretical and practical framework for translating these philosophical insights into ethical interaction policies that have the potential to impact many people. Our preference with this book has been to focus on the commonly accepted and valued concepts of beneficence and autonomy. Of course, given other theoretical commitments, we could have chosen different sets of concepts and values. But these concepts, in concert with the empirical evidence on the fragmentation of fundamental philosophical intuitions, uniquely inform some influential emerging debates with major policy impactions and direct relevance to people's daily lives (e.g., when and why might Libertarian Paternalism be the preferred method for helping people make decisions that promote health, wealth, and happiness?).

In this final chapter, we stretched to take what we think is an important step toward providing a well-grounded, simple, and sound approach to ethical interaction theory and interactive policy analysis. The common and overarching goal is obvious: Use systems and science to ethically and efficiently help people make better decisions—decisions that allow them to get more of what they (should and often do) value, including health, wealth, safety, and happiness. So that the intricate philosophical integration doesn't become lost for esoteric ends, we deliberately worked to give some structure to a prototype framework for heuristic evaluations of decision policies, following some best practices in systems design and human factors engineering fields. Ultimately, we are fairly comfortable with the notion that we've provided a rough but useful first draft of practical and easier-to-use tools for interactive policy analysis (e.g., evaluating decision policies ranging from paternalistic policies like nudges to representative education policies such as decision aids; see Appendix). However, we once again want to emphasize that we do not know (or think that we can know) all of the values that people should or do have. Thankfully, we don't need to know them all in order to adequately justify our approach because there is substantial agreement that autonomy is at least one essential and enduring human value, and for good reason. Consequently, provided that people have a representative understanding of a decision, helping people make their own decisions is likely to increase overall welfare in complex and surprising ways. These benefits result not only because autonomy provides opportunities to obtain those values, but also because it provides an essential and important route to individual self-efficacy and personal resiliency. Helping people become more autonomous has bountiful, measurable, and enduring well-being and welfare benefits (Deci & Ryan, 1995; Deci & Ryan, 2000; Devine, Camfield, & Gough, 2008; Ryan & Deci, 2017; Ryff, 1995; Sandman & Munthe, 2010) often giving rise to a deep sense of personal meaning and satisfaction (Peterson & Seligman, 2004; Seligman & Csikszentmihayli, 2000).

Beyond many more and less tangible benefits, ethical interaction theory provides an ethically defensible, evidence-based framework for the sustainable development of science for informed decision making—i.e., inclusive science designed to efficiently and ethically enhance skilled and autonomous decision making. After all, there is no question that providing domain-general education and decision making training can and often does help people make better decisions across a very wide range of high-stakes choices (Cokely et al., 2018; Garcia-Retamero et al., 2017). And while these interactions may often be about helping people obtain self-directed values (e.g., their own happiness, wealth, health), almost all humans have other-directed values, including values about entities and issues they may never even come in direct contact with (e.g., animals, organizations, future generations, biodiversity). Hence, at least on the average, it is a very good bet that helping people effectively realize and express their values should also make society better off independent of the specific benefits that accrue for the decision maker. In these ways, ethical interactive systems promote *autonomous* and *efficient beneficence for individuals and societies more generally*.

Given our analysis, a robust theoretically and empirically sound framework is now starting to come into place. More than many other decision science-based approaches, this framework provides for the protection of diverse values, choices, and individuals. On this view, decision scientists and policy makers have a duty to defend our freedom to disagree and to decide for ourselves. Our charge now is to make the most of these new resources and related insights from Ethical Interaction Theory.

Appendix: Heuristic Evaluation Checklist Example

Compare Relative Costs and Benefits

- Impact on autonomy:
  - Does any decision policy clearly infringe on autonomy?
  - Do current standards/benchmarks infringe on autonomy (e.g., the status quo)?
- Decision policy feasibility and implementation requirements:
  - Are development, production, and distributional costs similar of the compared policies?
  - Are maintenance and updating needs similar of the compared policies?

- General consumer risks and benefits:
  - What are the main cognitive and emotional impacts?
  - What are the main social, cultural, and environmental impacts?

    Influence on cooperation, communication, conflicts

  - What are the main costs and benefits to health, financial, social, and other outcomes, as compared to best alternatives and the status quo?
- Disparities and unequal impact:
  - Is there clear or potential conflicts of interest among targeted groups?
  - Does any decision policy promote controversial standards?
  - Are there differential influences or consequences for some individuals (e.g. minority groups, demographics, skill levels)?
- Robustness and durability
  - Is policy robust against distortion (e.g., unlikely to cause confusion)?
  - Is policy resilient to changes in socio-economic or political systems?
  - Will interventions be effective in routine and non-routine situations (e.g., effective in an emerging or during a crisis?)
- Risks to the choice architect
  - Does the policy promote trust, justice, and other social license to operate judgments?
  - What are the physical, emotional, reputation, and monetary risks to the choice architect?
- Net benefits (i.e., benefits—costs of each alternative):
  - Is the quality of the (scientific) evidence sufficient for fair comparison?

- Is there a clear benefit compared to standard benchmarks (status quo)?
- Is there a clear benefit compared to best available alternative?

References

Achinstein, P. (1994). Stronger evidence. *Philosophy of Science, 61*, 329–350.

Achinstein, P. (2000). Why philosophical theories of evidence are (and ought to be) ignored by scientists. *Philosophy of Science, 67*, S180–S192.

Adams, F. (1986). Intention and intentional action: The simple view. *Mind & Language, 1*, 281–301.

Adams, F., & Steadman, A. (2004a). Intentional action and moral considerations: Still pragmatic. *Analysis, 64*, 268–276.

Adams, F., & Steadman, A. (2004b). Intentional action in ordinary language: Core concept or pragmatic understanding? *Analysis, 64*, 173–181.

Akert, R. M., & Panter, A. T. (1988). Extraversion and the ability to decode nonverbal-communication. *Personality and Individual Differences, 9*(6), 965–972. https://doi.org/10.1016/0191-8869(88)90130-4

Alexander, J., & Weinberg, J. M. (2007). Analytic epistemology and experimental philosophy. *Philosophy Compass, 2*, 56–80.

Alexander, J., Betz, D., Gonnerman, C., & Waterman, J. P. (2018). Framing how we think about disagreement. *Philosophical Studies: An International Journal for Philosophy in the Analytic Tradition, 175*(10), 2539–2566. Retrieved from http://www.jstor.org/stable/45094063

Allan, J. N., Ripberger, J. T., Ybarra, V. T., & Cokely, E. T. (2017a). The Oklahoma Warning Awareness Scale: A psychometric evaluation of subjective awareness of natural hazard warnings. In *Proceedings of the Human Factors and Ergonomics Society Annual Meeting*.

A. Feltz, E. T. Cokely, *Diversity and Disagreement*,
https://doi.org/10.1007/978-3-031-61935-9

Allan, J. N., Ripberger, J. T., Ybarra, V. T., & Cokely, E. T. (2017b). Tornado risk literacy: Beliefs, biases, and vulnerability. In *Proceedings of the Naturalistic Decision Making 13th International Conference.*

American Psychological Association. (2013). *Diagnostic and statistical manual of mental disorders* (5th ed.). American Psychiatric Publishing.

Anderson, B. L., & Schulkin, J. (Eds.). (2014). *Numerical reasoning in judgments and decision making about health.* Cambridge University Press.

Anderson, J. R. (1991). The adaptive nature of human categorization. *Psychological Review, 98*(3), 409–429. https://doi.org/10.1037/0033-295X.98.3.409

Andow, J., & Cova, F. (2016). Why compatibilist intuitions are not mistaken: A reply to Feltz and Millan. *Philosophical Psychology, 29*, 550–566.

Anscombe, G. E. M. (1958). Modern moral philosophy. *Philosophy, 33*, 1–19.

Appiah, A. (2008). *Experiments in ethics.* Harvard University Press.

Archard, D. (2011). Why moral philosophers are not and should not be moral experts. *Bioethics, 25*(3), 119–127. https://doi.org/10.1111/j.1467-8519.2009.01748.x

Aristotle, Ross, W. D., & Urmson, J. O. (1980). *The Nicomachean ethics.* Oxford University Press.

Arpaly, N. (2003). *Unprincipled virtue: An inquiry into moral agency.* Oxford University Press.

Ashton, M. C., Lee, K., & Paunonen, S. V. (2002). What is the central feature of extraversion? Social attention versus reward sensitivity. *Journal of Personality and Social Psychology, 83*(1), 245–252. https://doi.org/10.1037//0022-3514.83.1.245

Ashton, M. C., Lee, K., de Vries, R. E., Hendrickse, J., & Born, M. P. H. (2012). The maladaptive personality traits of the Personality Inventory for DSM-5 (PID-5) in relation to the HEXACO personality factors and schizotypy/dissociation. *Journal of Personality Disorders, 26*(5), 641–659. https://doi.org/10.1521/pedi.2012.26.5.641

Audi, R. (2011). *Epistemology: A contemporary introduction to the theory of knowledge* (3rd ed.). Routledge.

Ayer, A. J. (1952). *Language, truth, and logic.* Dover Publications.

Ayer, A. J. (1954). *Philosophical essays.* Macmillan; St. Martin's Press.

Bandura, A. (1986). *Social foundations of thought and action: A social cognitive theory.* Prentice-Hall.

Baron, J. (1978). Intelligence and general strategies. In G. Underwood (Ed.), *Strategies of information processing* (pp. 403–450). Academic Press.

Baron, J. (1985). *Rationality and intelligence.* Cambridge University Press.

Baron, J. (2008). *Thinking and deciding* (4th ed.). Cambridge University Press.

Baron, R. A., & Branscombe, N. R. (2012). *Social psychology* (13th ed.). Pearson.

Barton, A., Cokely, E. T., Galesic, M., Koehler, A., & Haas, M. (2009). Comparing risk reductions: On the dynamic interplay of cognitive strategies, numeracy, complexity, and format. In N. A. Taatgen & H. van Rijn (Eds.), *Proceedings of the 31st Annual Conference of the Cognitive Science Society* (pp. 2347–2352).

Baumeister, R. F. (2008). Free will in scientific psychology. *Perspectives on Psychological Science, 3*, 14–19.

Baumeister, R. F., Sparks, E. A., Stillman, T. F., & Vohs, K. D. (2008). Free will in consumer behavior: Self-control, ego depletion, and choice. *Journal of Consumer Psychology, 18*(1), 4–13. https://doi.org/10.1016/J.Jcps.2007.10.002

Baumeister, R. F., Masicampo, E. J., & DeWall, C. N. (2009). Prosocial benefits of feeling free: Disbelief in free will increases aggression and reduces helpfulness. *Personality and Social Psychology Bulletin, 35*(2), 260–268. https://doi.org/10.1177/0146167208327217

Bealer, G. (1998). Intuition and the autonomy of philosophy. In M. DePaul & W. Ramsey (Eds.), *Rethinking intuition: The psychology of intuition and its role in philosophical inquiry* (pp. 201–239). Rowman and Littlefield.

Beauchamp, T. L. (2003). The nature of applied ethics. In R. G. Frey & C. H. Wellman (Eds.), *A companion to applied ethics* (pp. 1–16). Blackwell Publishing.

Beauchamp, T. L., & Childress, J. F. (2009). *Principles of biomedical ethics* (6th ed.). Oxford University Press.

Beebe, J. R., & Sackris, D. (2016). Moral objectivism across the lifespan. *Philosophical Psychology, 29*(6), 912–929. https://doi.org/10.1080/09515089.2016.1174843

Benartzi, S., Beshears, J., Milkman, K., Sunstein, C. R., Thaler, R. H., Shankar, M., et al. (2017). Should governments invest more in nudging? *Psychological Science, 28*, 1–15.

Benn, S. (1976). Freedom, autonomy and the concept of a person. *Proceedings of the Aristotelian Society, 76*, 109–130.

Bentham, J. (2008). *An introduction to the principles of morals and legislation.* Barnes & Noble, Inc.

Berleur, J., Nurminen, M. I., & Impagliazzo, J. (2006). *Social informatics: An information society for all?: In remembrance of Rob Kling: Proceedings of the Seventh International Conference on Human Choice and Computers (HCC7), IFIP TC 9, Maribor, Slovenia, September 21–23, 2006.* Springer.

Bernstein, M. (2002). Fatalism. In R. Kane (Ed.), *The Oxford handbook of free will* (pp. 65–81). Oxford University Press.

Bishop, M. A. (2015). *The good life: Unifying the philosophy and psychology of well-being.* Oxford University Press.

Bishop, M. A., & Trout, J. D. (2005). *Epistemology and the psychology of human judgment.* Oxford University Press.

Bloom, B. (1985). *Developing talent in young people.* Random House.

Blumenthal-Barby, J. S., & Burroughs, H. (2012). Seeking better health care outcomes: The ethics of using the "nudge". *American Journal of Bioethics, 12*(2), 1–10. https://doi.org/10.1080/15265161.2011.634481

Boardman, A. E., Greenberg, D. H., Vining, A. R., & Weimer, D. L. (2011). *Cost-Benefit Analysis: Concepts and Practice* (4th ed.). Upper Saddle River, NJ: Prentice Hall.

Bonjour, L. (1998). *In defense of pure reason: A rationalist account of a priori justification.* Cambridge University Press.

Bouchard, T. J., Jr., & McGue, M. (2003). Genetic and environmental influences on human psychological differences. *Journal of Neurobiology, 54*(1), 4–45. https://doi.org/10.1002/neu.10160

Bouchard, T. J., & Loehlin, J. C. (2001). Genes, evolution, and personality. *Behavior Genetics, 31*(3), 243–273.

Bourget, D., & Chalmers, D. (2014). What do philosophers believe? *Philosophical Studies, 170*, 465–500.

Brady, M. S. (2005). The value of the virtues. *Philosophical Studies, 125*(1), 85–113. https://doi.org/10.1007/S11098-004-7788-Z

Bratman, M. (1984). Two faces of intention. *The Philosophical Review, 93*, 375–405.

Broad, C. D. (1952). *Ethics and the history of philosophy; selected essays.* Routledge & K. Paul.

Buchanan, A. (1978). Medical paternalism. *Philosophy & Public Affairs, 7*(4), 370–390. Retrieved from http://www.ncbi.nlm.nih.gov/pubmed/11664929

Buchanan, A. E., & Brock, D. W. (1989). *Deciding for others: The ethics of surrogate decision making.* Cambridge University Press.

Buckwalter, W., & Stich, S. (2014). Gender and philosophical intuition. In J. Knobe & S. Nichols (Eds.), *Experimental philosophy* (Vol. 2, pp. 307–346). Oxford University Press.

Calder, T. (2007). Against consequentialist theories of virtue and vices. *Utilitas, 19*, 201–219.

Cappelen, H. (2012). *Philosophy without intuitions* (1st ed.). Oxford University Press.

Cawley, M. J., Martin, J. E., & Johnson, J. A. (2000). A virtues approach to personality. *Personality and Individual Differences, 28*(5), 997–1013. https://doi.org/10.1016/S0191-8869(99)00207-X

Chamorro-Premuzic, T., Furnham, A., & Ackerman, P. L. (2006). Ability and personality correlates of general knowledge. *Personality and Individual Differences, 41*(3), 419–429. https://doi.org/10.1016/J.Paid.2005.11.036

Chang, R. (1997). Introduction. In R. Change (Ed.), *Incommensurability, incomparability, and practical reason* (pp. 1–34). Harvard University Press.

Chang, W., & Tetlock, P. E. (2016). Rethinking the training of intelligence analysts. *Intelligence and National Security, 31*(6), 903–920.

Chang, W., Chen, E., Mellers, B., & Tetlock, P. (2016). Developing expert political judgment: The impact of training and practice on judgmental accuracy in geopolitical forecasting tournaments. *Judgment and Decision Making, 11*(5), 509–526.

Cho, J., Cokely, E. T., Ramasubramanian, M., Allan, J. N., Feltz, A., & Garcia-Retamero, R. (2024). Numeracy does not polarize climate change judgments: Numerate people are more knowledgeable and knowledge is power. *Decision, 11*(2), 320–344. https://doi.org/10.1037/dec0000223

Claxton, G. (1998). Investigating human intuition: Knowing without knowing why. *The Psychologist, 11*, 217–220.

Cohen, L. J. (1981). Investigating human intuition: Knowing without knowing why. *The Behavioral and Brain Sciences, 4*, 317–331.

Cokely, E. T. (2009). Beyond generic dual processes: How should we evaluate scientific progress? *PsycCritiques, 54*(51), 10. https://doi.org/10.1037/a0017643

Cokely, E. T., & Feltz, A. (2009a). Adaptive variation in judgment and philosophical intuition Reply. *Consciousness and Cognition, 18*(1), 356–358. https://doi.org/10.1016/J.Concog.2009.01.001

Cokely, E. T., & Feltz, A. (2009b). Individual differences, judgment biases, and theory-of-mind: Deconstructing the intentional action side effect asymmetry. *Journal of Research in Personality, 43*(1), 18–24. https://doi.org/10.1016/j.jrp.2008.10.007

Cokely, E. T., & Feltz, A. (2011). Virtue in business: Morally better, praiseworthy, trustworthy, and more satisfying. *Journal of Organizational Moral Psychology, 2*, 13–26.

Cokely, E. T., & Kelley, C. M. (2009). Cognitive abilities and superior decision making under risk: A protocol analysis and process model evaluation. *Judgment and Decision Making, 4*(1), 20–33.

Cokely, E. T., Kelley, C. M., & Gilchrist, A. L. (2006). Sources of individual differences in working memory: Contributions of strategy to capacity. *Psychonomic Bulletin & Review, 13*(6), 991–997. https://doi.org/10.3758/Bf03213914

Cokely, E. T., Parpart, P., & Schooler, L. J. (2009). On the link between cognitive control and heuristic processes. In N. A. Taatgen & H. van Rijn (Eds.), *Proceedings of the 31st Annual Conference of the Cognitive Science Society* (pp. 2926–2931). Cognitive Science Society.

Cokely, E. T., Galesic, M., Schulz, E., Ghazal, S., & Garcia-Retamero, R. (2012). Measuring risk literacy: The Berlin numeracy test. *Judgment and Decision Making, 7*(1), 25–47.

Cokely, E. T., Ghazal, S., Garcia-Retamero, R., & Galesic, M. (2013). How to measure risk literacy in educated samples. In R. Garcia-Retamero & M. Galesic (Eds.), *Transparent communication of risks about health: Overcoming cultural differences* (pp. 29–52). Springer.

Cokely, E. T., Ghazal, S., & Garcia-Retamero, R. (2014). Measuring numeracy. In B. L. Anderson & J. Schulkin (Eds.), *Numerical reasoning in judgment and decision making about health* (pp. 11–38). Cambridge University Press.

Cokely, E. T., Feltz, A., Ghazal, S., Allan, J., Petrova, D., & Garcia-Retamero, R. (2018). Skilled decision theory: From intelligence to numeracy and expertise. In A. Ericsson, R. Hoffman, A. Kozbelt, & A. Williams (Eds.), *Cambridge handbook of expertise and expert performance* (pp. 476–505). Cambridge University Press.

Cokely, E. T., Cho, J., Allan, J. N., Ramasubramanian, M., Feltz, A., & Garcia-Retamero, R. (in preparation). Risk literacy knowledge is power: On the dominant role of representative understanding in skilled decision making. In M. Williams, A. Kozbelt, F. Preckel, & R. Hoffman (Eds.), *The Cambridge handbook of expertise and expert performance*. Cambridge University Press.

Cooke, D. J., Forth, A. E., & Hare, R. D. (1998). *Psychopathy: Theory, research, and implications for society*. Kluwer Academic.

Copp, D., & Sobel, D. (2004). Morality and virtue: An assessment of some recent work in virtue ethics. *Ethics, 114*(3), 514–554. https://doi.org/10.1086/382058

Costa, P. T., & McCrae, R. R. (1992). The five-factor model of personality and its relevance to personality disorders. *Journal of Personality Disorders, 6*(4), 343–359. https://doi.org/10.1521/pedi.1992.6.4.343

Costa, P. T., & Mccrae, R. R. (1988). From catalog to classification—Murray needs and the 5-factor model. *Journal of Personality and Social Psychology, 55*(2), 258–265. https://doi.org/10.1037//0022-3514.55.2.258

Cova, F., Bertoux, M., Bourgeois-Gironde, S., & Dubois, B. (2012). Judgments about moral responsibility and determinism in patients with behavioural variant of frontotemporal dementia: Still compatibilists. *Consciousness and Cognition, 21*(2), 851–864. https://doi.org/10.1016/J.Concog.2012.02.004

Cova, F., Strickland, B., Abatista, A., Allard, A., Andow, J., Attie, M., Beebe, J., Berniūnas, R., Boudesseul, J., Colombo, M., Cushman, F., Diaz, R., N'Djaye Nikolai van Dongen, N., Dranseika, V., Earp, B. D., Torres, A. G., Hannikainen, I., Hernández-Conde, J. V., Hu, W., et al. (2021). Estimating the reproducibility of experimental philosophy. *Review of Philosophy and Psychology, 12*(1), 9–44. https://doi.org/10.1007/s13164-018-0400-9. (Retraction published 2021, Review of Philosophy and Psychology, 12[1], 45–48).

Crisp, R. (2010). Virtue ethics and virtue epistemology. *Metaphilosophy, 41*(1–2), 22–40. Retrieved from <Go to ISI>://000273317000002.

Crosthwaite, J. (1995). Moral expertise: A problem in the professional ethics of professional ethicists. *Bioethics, 9*(5), 361–379. Retrieved from https://www.ncbi.nlm.nih.gov/pubmed/11653255

Cumming, G., & Calin-Jageman, R. (2017). *Introduction to the new statistics: Estimation, open science, and beyond*. Routledge.

Cummins, R. (1998). Reflection on reflective equilibrium. In M. DePaul & W. Ramsey (Eds.), *Rethinking intuition: The psychology of intuition and its role in philosophical inquiry* (pp. 113–127). Rowman & Littlefield Publishers.

Cushman, F., & Mele, A. (2008). Intentional action: Two-and-a-half folk concepts? In J. Knobe & S. Nichols (Eds.), *Experimental philosophy* (pp. 171–188). Oxford University Press.

Daniels, N. (1979). Wide reflective equilibrium and theory acceptance in ethics. *Journal of Philosophy, 76*, 256–282.

Davidson, D. (1987). Knowing one's own mind. *Proceedings and Addresses of the American Philosophical Association, 61*, 441–458.

De Brigard, F., Mandelbaum, E., & Ripley, D. (2009). Responsibility and the brain sciences. *Ethical Theory and Moral Practice, 12*, 511–524.

Deci, E. L., & Ryan, R. M. (1995). Human autonomy. In M. H. Kernis (Ed.), *Efficacy, agency, and self-esteem* (pp. 31–49). Springer.

Deci, E. L., & Ryan, R. M. (2000). The "what" and "why" of goal pursuits: Human needs and the self-determination of behavior. *Psychological Inquiry, 11*, 227–268.

Dennett, D. C. (1984). *Elbow room: The varieties of free will worth wanting*. Clarendon Press; Oxford University Press.

Deutsch, M. (2010). Intuitions, counter-examples, and experimental philosophy. *Review of Philosophy and Psychology, 1*, 447–460.

Deutsch, M. (2015). *The myth of the intuitive: Experimental philosophy and philosophical method*. The MIT Press, a Bradford Book.

Devine, J., Camfield, L., & Gough, I. (2008). Autonomy or dependence—Or both?: Perspective from Bangladesh. *Journal of Happiness Studies, 9*, 105–138.

Devitt, M. (2006). Intuitions in linguistics. *British Journal for the Philosophy of Science, 57*, 481–513.

Diener, E., Emmons, R. A., Larsen, R. J., & Griffin, S. (1985). The satisfaction with life scale. *Journal of Personality Assessment, 49*(1), 71–75. https://doi.org/10.1207/S15327752jpa4901_13

DiPazz, S., & Eccles, R. (2002). *Building public trust: The future of corporate reporting*. Wiley.

Dongen, N., Colombo, M., Romero, F., & Sprenger, J. (2020). Intuitions about the reference of proper names: A meta-analysis. *The Review of Philosophy and Psychology, 12*, 745–774.

Doris, J. M. (2002). *Lack of character: Personality and moral behavior*. Cambridge University Press.

Drane, J. F. (1985). The many faces of competency. *Hastings Center Report, 15*(2), 17–21. https://doi.org/10.2307/3560639

Driver, J. (1995). Monkeying with motives: Agent-basing virtue ethics. *Utilitas, 7*, 281–288.

Driver, J. (2001). *Uneasy virtue*. Cambridge University Press.

Driver, J. (2003). The conflation of moral and epistemic virtue. *Metaphilosophy, 34*, 367–383.

Driver, J. (2004). Response to my critics. *Utilitas, 16*, 33–41.

Dworkin, G. (1981). The concept of autonomy. *Grazer Philosophische Studien, 12*, 203–213.

Dworkin, G. (1988). *The theory and practice of autonomy*. Cambridge University Press.

Ekstrom, L. W. (2002). Libertarianism and Frankfurt-style cases. In R. Kane (Ed.), *The Oxford handbook of free will* (pp. 309–322). Oxford University Press.

Ellis, S. (2008). The main argument for value incommensurability (and why it fails). *The Southern Journal of Philosophy, 46*, 27–43.

Ellis, K. M., Cokely, E. T., Ghazal, S., & Garcia-Retamero, R. (2014). Do people understand their home HIV test results? Risk literacy and information search. *Proceedings of the Human Factors and Ergonomics Society Annual Meeting, 58*(1), 1323–1327. SAGE Publications.

Ericsson, K. A., & Kintsch, W. (1995). Long-term working-memory. *Psychological Review, 102*, 211–245.

Ericsson, K. A., & Lehmann, A. (1996). Expert and exceptional performance: Evidence of maximal adaptions to task constraints. *Annual Review of Psychology, 47*, 273–305.

Ericsson, K. A., & Simon, H. A. (1984). *Protocol analysis: Verbal reports as data*. MIT Press.

Ericsson, K. A., Charness, N., Feltovich, P. J., & Hoffman, R. R. (Eds.). (2006). *The Cambridge handbook of expertise and expert performance*. Cambridge University Press.

Ericsson, K. A., Prietula, M. J., & Cokely, E. T. (2007). The making of an expert. *Harvard Business Review, 85*(7–8), 114–+.

Ericsson, K. A., Hoffman, R. R., Kozbelt, A., & Williams, A. M. (Eds.). (2018). *The Cambridge handbook of expertise and expert performance* (2nd ed.). Cambridge: Cambridge University Press.

Estes, W. (1956). The problem of inference from curves based on group data. *Psychological Bulletin, 53*, 134–140.

Evans, J., & Frankish, K. (2009). *In two minds: Dual processes and beyond*. Oxford University Press.

Feinberg, J. (1986). *Harm to self*. Oxford University Press.

Felsen, G., Castelo, N., & Reiner, P. (2013). Decisional enhancement and autonomy: Public attitudes toward overt and covert nudges. *Judgment and Decision Making, 8*, 202–213.

Feltz, A. (2008). Problems with the appeal to intuition in epistemology. *Philosophical Explorations, 11*, 131–141.

Feltz, A. (2013). Pereboom and premises: Asking the right questions in the experimental philosophy of free will. *Consciousness and Cognition, 22*(1), 53–63. https://doi.org/10.1016/j.concog.2012.11.007

Feltz, A. (2015a). Ethical information transparency and sexually transmitted diseases. *Current HIV Research, 13*, 421–431.

Feltz, A. (2015b). Experimental philosophy of actual and counterfactual free will intuitions. *Consciousness and Cognition: An International Journal, 36*, 113–130. https://doi.org/10.1016/j.concog.2015.06.001

Feltz, A., & Bishop, M. (2010). The proper role of intuitions in epistemology. In M. Milkowski & K. Talmont-Kaminiski (Eds.), *Beyond description: Naturalism and normativity* (pp. 101–122). College Publications.

Feltz, A., & Cokely, E. T. (2007). An anomaly in intentional action ascriptions: More evidence of folk diversity. In D. S. McNamara & J. G. Trafton (Eds.), *Proceedings of the 29th Annual Cognitive Science Society* (p. 1748). Cognitive Science Society.

Feltz, A., & Cokely, E. T. (2008). The fragmented folk: More evidence of stable individual differences in moral judgments and folk intuitions. In B. C. Love, K. McRae, & V. M. Sloutsky (Eds.), *Proceedings of the 30th Annual Conference of the Cognitive Science Society* (pp. 1771–1776). Cognitive Science Society.

Feltz, A., & Cokely, E. T. (2009). Do judgments about freedom and responsibility depend on who you are? Personality differences in intuitions about compatibilism and incompatibilism. *Consciousness and Cognition, 18*(1), 342–350. https://doi.org/10.1016/j.concog.2008.08.001

Feltz, A., & Cokely, E. T. (2011). Individual differences in theory-of-mind judgments: Order effects and side effects. *Philosophical Psychology, 24*, 343–355.

Feltz, A., & Cokely, E. T. (2012a). The philosophical personality argument. *Philosophical Studies, 161*(2), 227–246. https://doi.org/10.1007/s11098-011-9731-4

Feltz, A., & Cokely, E. T. (2012b). The virtues of ignorance. *The Review of Philosophy and Psychology, 3*, 335–350.

Feltz, A., & Cokely, E. T. (2013a). Predicting philosophical disagreement. *Philosophy Compass, 8*, 978–989.

Feltz, A., & Cokely, E. T. (2013b). Virtue or consequences: The folk against pure evaluational internalism. *Philosophical Psychology, 26*(5), 702–717. https://doi.org/10.1080/09515089.2012.692903

Feltz, A., & Cokely, E. T. (2016). Personality and philosophical bias. In J. Sytsma & W. Buckwalter (Eds.), *A companion to experimental philosophy* (pp. 578–589). Wiley.

Feltz, A., & Cokely, E. T. (2019). Extraversion and compatibilist intuitions: A ten-year retrospective and meta-analysis. *Philosophical Psychology, 32*(3), 388–403.

Feltz, A., & Cova, F. (2014). Moral responsibility and free will: A meta-analysis. *Consciousness and Cognition, 30*, 234–246. https://doi.org/10.1016/j.concog.2014.08.012

Feltz, S., & Feltz, A. (2019). The knowledge of animal as food scale. *Human-Animal Interaction Bulletin, 7*(2), 19–45.

Feltz, A., & Millan, M. (2015). An error theory for compatibilist intuitions. *Philosophical Psychology, 28*(4), 529–555.

Feltz, A., Cokely, E. T., & Nadelhoffer, T. (2009). Natural compatibilism versus natural incompatibilism: Back to the drawing board. *Mind & Language, 24*(1), 1–23. https://doi.org/10.1111/J.1468-0017.2008.01351.X

Feltz, A., Harris, M., & Perez, A. (2012a). Perspective in intentional action attribution. *Philosophical Psychology, 25*, 335–350.

Feltz, A., Perez, A., & Harris, M. (2012b). Free will causes, and decisions individual differences in written reports. *Journal of Consciousness Studies, 19*(9-10), 166–189. Retrieved from <Go to ISI>://WOS:000319628200008.

Feltz, A., Tanner, B., Hoang, G., Holt, J., & Muhammad, A. (2022). Free will and skilled decision theory. In T. Nadelhoffer & A. Monroe (Eds.), *Advances in experimental philosophy of free will and responsibility* (pp. 185–202). Bloomsbury.

Fischer, J. M., & Ravizza, M. (1998). *Responsibility and control: A theory of moral responsibility.* Cambridge University Press.

Flanagan, O. (1990). Virtue and ignorance. *Journal of Philosophy, 87*(8), 420–428. https://doi.org/10.2307/2026736

Flanagan, O. (1991). *Varieties of moral personality: Ethics and psychological realism.* Harvard University Press.

Flavell, J. H. (1979). Metacognition and cognitive monitoring: A new area of cognitive-developmental inquiry. *American Psychologist, 34*, 906–911.

Foot, P. (2001). *Natural goodness.* Oxford University Press.

Fox, M. C., Ericsson, K. A., & Best, R. (2011). Do procedures for verbal reporting of thinking have to be reactive? A meta-analysis and recommendations for best reporting methods. *Psychological Bulletin, 137*(2), 316.

Frankfurt, H. (1969). Alternative possibilities and moral responsibility. *The Journal of Philosophy, 66*, 829–839.

Frankfurt, H. (1971). Freedom of the will and the concept of a person. *The Journal of Philosophy, 68*, 5–20.

Frederick, S. (2005). Cognitive reflection and decision making. *Journal of Economic Perspectives, 19*(4), 25–42. https://doi.org/10.1257/089533005775196732

Frederick, S. (2008). IQ/cognitive reflection, risk taking and the role of task instructions. *International Journal of Psychology, 43*(3-4), 12–12. Retrieved from <Go to ISI>://000259264300132.

Funder, D. C. (1991). Global traits—A neo-allportian approach to personality. *Psychological Science, 2*(1), 31–39. https://doi.org/10.1111/J.1467-9280.1991.Tb00093.X

Funder, D. C. (1995). On the accuracy of personality judgment—A realistic approach. *Psychological Review, 102*(4), 652–670. https://doi.org/10.1037//0033-295x.102.4.652

Funder, D. C. (2001). Personality. *Annual Review of Psychology, 52*, 197–221.

Galesic, M., & Garcia-Retamero, R. (2011). Graph literacy: A cross-cultural comparison. *Medical Decision Making, 31*(3), 444–457. https://doi.org/10.1177/0272989X10373805

Garcia-Retamero, R., & Cokely, E. T. (2012). Advances in efficient health communication: Promoting prevention and detection of STDs. *Current HIV Research, 10*(3), 262–270.

Garcia-Retamero, R., & Cokely, E. T. (2013a). Reducing the effect of framed messages about health. In R. Garcia-Retamero & M. Galesic (Eds.), *Transparent communication of risks about health: Overcoming cultural differences* (pp. 165–191). Springer.

Garcia-Retamero, R., & Cokely, E. T. (2013b). Communicating health risks with visual aids. *Current Directions in Psychological Science, 22*(5), 392–399. https://doi.org/10.1177/0963721413491570

Garcia-Retamero, R., & Cokely, E. T. (2013c). The influence of skills, message frame, and visual aids on prevention of sexually transmitted diseases. *Journal of Behavioral Decision Making, 27*(2), 179–189.

Garcia-Retamero, R., & Cokely, E. T. (2014a). The influence of skills, message frame, and visual aids on prevention of sexually transmitted diseases. *Journal of Behavioral Decision Making, 27*(2), 179–189.

Garcia-Retamero, R., & Cokely, E. T. (2014b). Using visual aids to help people with low numeracy make better decisions. In B. L. Anderson & J. Schulkin (Eds.), *Numerical reasoning in judgment and decision making about health* (pp. 153–174). Cambridge University Press.

Garcia-Retamero, R., & Cokely, E. T. (2015a). Simple but powerful health messages for increasing condom use in young adults. *The Journal of Sex Research, 52*(1), 30–42. https://doi.org/10.1080/00224499.2013.806647

Garcia-Retamero, R., & Cokely, E. T. (2015b). Brief messages to promote prevention and detection of sexually transmitted infections. *Current HIV Research, 13*(5), 408–420.

Garcia-Retamero, R., & Cokely, E. T. (2017). Designing visual aids that promote risk literacy: A systematic review of health research and evidence-based heuristics. *Human Factors, 59*(4), 582–627.

Garcia-Retamero, R., Okan, Y., & Cokely, E. T. (2012). Using visual aids to improve communication of risks about health: A review. *ScientificWorldJournal, 2012*, 562637. https://doi.org/10.1100/2012/562637

Garcia-Retamero, R., Wicki, B., Cokely, E. T., & Hanson, B. (2014). Factors predicting surgeons' preferred and actual roles in interactions with their patients. *Health Psychology, 33*(8), 920.

Garcia-Retamero, R., Cokely, E. T., & Hoffrage, U. (2015). Visual aids improve diagnostic inferences and metacognitive judgment calibration. *Frontiers in Psychology, 6*, 932. https://doi.org/10.3389/fpsyg.2015.00932

Garcia-Retamero, R., Cokely, E. T., Ghazal, S., & Joeris, A. (2016a). Measuring graph literacy without a test: A brief subjective assessment. *Medical Decision Making, 36*(7), 854–867.

Garcia-Retamero, R., Cokely, E. T., Wicki, B., & Joeris, A. (2016b). Improving risk literacy in surgeons. *Patient Education and Counseling, 99*(7), 1156–1161.

Garcia-Retamero, R., Petrova, D., Feltz, A., & Cokely, E. T. (2017). *Measuring Graph Literacy*. Oxford Research Encyclopedia of Communication. Retrieved 11 Mar. 2024, from https://oxfordre.com/communication/view/10.1093/acrefore/9780190228613.001.0001/acrefore-9780190228613-e-302.

Garcia-Retamero, R., Petrova, D., Cokely, E. T., & Joeris, A. (2019a). Scientific risk reporting in medical journals can bias expert judgment: Comparing surgeons' risk comprehension across reporting formats. *Journal of Experimental Psychology: Applied, 26*(2), 283.

Garcia-Retamero, R., Sobkow, A., Petrova, D., Garrido, D., & Traczyk, J. (2019b). Numeracy and risk literacy: What have we learned so far? *The Spanish Journal of Psychology, 22*, E10.

Garrido, D., Petrova, D., Cokely, E., Carballo, G., & Garcia-Retamero, R. (2021). Parental risk literacy is related to quality of life in Spanish families of children with autism spectrum disorder. *Journal of Autism and Developmental Disorders, 51*(7), 2475–2484.

Gershoff, A. D., & Koehler, J. J. (2011). Safety first? The role of emotion in safety product betrayal aversion. *Journal of Consumer Research, 38*(1), 140–150. https://doi.org/10.1086/658883

Gert, B., & Culver, C. M. (1979). Justification of paternalism. *Ethics, 89*(2), 199–210. https://doi.org/10.1086/292097

Ghazal, S., Cokely, E. T., & Garcia-Retamero, R. (2014). Predicting biases in very highly educated samples: Numeracy and metacognition. *Judgment and Decision Making, 9*(1), 15.

Gigerenzer, G. (2001). The adaptive toolbox. In G. Gigerenzer & R. Selten (Eds.), *Bounded Rationality: The Adaptive Toolbox* (pp. 37–50). The MIT Press.

Gigerenzer, G. (2006). Heuristics. In G. Gigerenzer & C. Engel (Eds.), *Heuristics and the Law* (pp. 17–44). MIT Press; Dahlem University Press.

Gigerenzer, G. (2008). Why heuristics work. *Perspectives on Psychological Science, 3*(1), 20–29.

Gigerenzer, G. (2010). Moral satisficing: Rethinking moral behavior as bounded rationality. *Topics in Cognitive Science, 2*(3), 528–554. Retrieved from <Go to ISI>://WOS:000283869500016.

Gigerenzer, G., & Brighton, H. (2009). Homo heuristicus: why biased minds make better inferences. *Topics in Cognitive Science, 1*(1), 107–143. https://doi.org/10.1111/j.1756-8765.2008.01006.x

Gigerenzer, G., & Goldstein, D. G. (1996). Reasoning the fast and frugal way: Models of bounded rationality. *Psychological Review, 103*(4), 650.

Gigerenzer, G., & Regier, T. (1996). How do we tell an association from a rule? Comment on Sloman. *Psychological Bulletin, 119*(1), 23–26. https://doi.org/10.1037/0033-2909.119.1.23

Gigerenzer, G., & Selten, R. (2001). Rethinking rationality. In *Bounded rationality: The adaptive toolbox* (Vol. 1, p. 12).

Gigerenzer, G., Todd, P. M., & The ABC Research Group. (1999). *Simple heuristics that make us smart.* Oxford University Press.

Goldman, A. I., & Pust, J. (1998). Philosophical theory and intuitional evidence. In M. DePaul & W. Ramsey (Eds.), *Rethinking intuition: The psychology of intuition and its role in philosophical inquiry* (pp. 291–309). Rowman and Littlefield.

Goodman, N. (1955). *Fact, fiction, and forecast.* Bobbs-Merrill Company.

Goodwin, G. P., & Darley, J. M. (2006). The psychology of meta-ethics. Exploring objectivism. *Cognition, 106*, 1339–1366.

Gosling, S. D., Rentfrow, P. J., & Swann, W. B. (2003). A very brief measure of the Big-Five personality domains. *Journal of Research in Personality, 37*(6), 504–528. https://doi.org/10.1016/S0092-6566(03)00046-1

Gramlich, E. M. (1990). *A guide to benefit-cost analysis* (2. ed.). Englewood Cliffs, NJ: Prentice Hall.

Grice, H. P. (1975). Logic and conversation. In P. Cole & J. L. Morgan (Eds.), *Syntax and semantics* (pp. 41–58). Academic Press.

Haidt, J., & Graham, J. (2007). When morality opposes justice: Conservatives have moral intuitions that liberals may not recognize. *Social Justice Research, 20*, 98–116.

Haidt, J., Koller, S. H., & Dias, M. G. (1993). Affect, culture, and morality, or is it wrong to eat your dog. *Journal of Personality and Social Psychology, 65*(4), 613–628. https://doi.org/10.1037/0022-3514.65.4.613

Hales, S. (2000). The problem of intuition. *American Philosophical Quarterly, 37*, 125–147.

Hammond, K. R. (2000). Coherence and correspondence theories in judgment and decision making. In T. Connolly & H. R. Arkes (Eds.), *Judgment and decision making: An interdisciplinary reader* (2nd ed., pp. 53–65). Cambridge University Press.

Hansen, P., & Jespersen, A. (2013). Nudge and the manipulation of choice: A framework for the responsible use of the nudge approach to behaviour change in public policy. *European Journal of Risk Regulation, 3*, 3–28.

Hardman, D., & Macchi, L. (2003). *Thinking: Psychological perspectives on reasoning, judgment, and decision making.* Wiley.

Hare, R. M. (1989). Why do applied ethics? In R. M. Hare (Ed.), *Essays in ethical theory* (pp. 1–13). Clarendon Press.

Harman, G. (1999). Moral philosophy meets social psychology: Virtue ethics and the fundamental attribution error. *Proceedings of the Aristotelian Society, 99*, 315–331.

Hastie, R. (2001). Problems for judgment and decision making. *Annual Review of Psychology, 52*, 653–853.

Hausman, D. M., & Welch, B. (2010). Debate: To nudge or not to nudge. *Journal of Political Philosophy, 18*(1), 123–136. https://doi.org/10.1111/J.1467-9760.2009.00351.X

Haworth, L. (1986). *Autonomy: An essay in philosophical psychology and ethics.* Yale University Press.

Hertwig, R., & Grune-Yanoff, T. (2017). Nudging and boosting: Steering or empowering good decisions. *Perspectives on Psychological Science, 12*(6), 973–986. https://doi.org/10.1177/1745691617702496

Hoang, U., Feltz, S., Offer-Westort, T., & Feltz, A. (2023). Willingness to consume fewer animal products: A latent profile analysis. *Anthrozoos, 36*, 641–663. https://doi.org/10.1080/08927936.2023.2204640

Hooker, B. (2002). The collapse of virtue ethics. *Utilitas, 14*, 22–40.

Horne, Z., & Livengood, J. (2017). Ordering effects, updating effects, and the specter of global skepticism. *Synthese, 194*(4), 1189–1218. Retrieved from http://www.jstor.org/stable/26166392

Horvath, J. (2010). How (not) to react to experimental philosophy. *Philosophical Psychology, 23*, 448–480.

Horvath, J. (2015). Thought experiments and experimental philosophy. In C. Daly (Ed.), *The Palgrave handbook of philosophical methods* (pp. 386–418). Palgrave Macmillan.

Horvath, J., & Wiegmann, A. (2022). Intuitive expertise in moral judgments. *Australasian Journal of Philosophy, 100*(2), 342–359. https://doi.org/10.1080/00048402.2021.1890162

Huebner, B., Bruno, M., & Sarkissian, H. (2010). What does the nation of China think about phenomenal states? *Review of Philosophy and Psychology, 1*, 255–243.

Hume, D., Selby-Bigge, L. A., & Nidditch, P. H. (1978). *A treatise of human nature* (2nd ed.). Clarendon Press; Oxford University Press.

Hurka, T. (2006). Virtuous acts, virtuous dispositions. *Analysis, 66*, 69–76.

Hurka, T. (2010). Right act, virtuous dispositions. *Metaphilosophy, 41*, 58–72.

Hursthouse, R. (1999). *On virtue ethics.* Oxford University Press.

Jackson, F. (1998). *From metaphysics to ethics: A defence of conceptual analysis.* Oxford University Press.

Jacoby, L. L., Kelley, C. M., & McElree, B. D. (1999). The role of cognitive control: Early selection versus late correction. In S. Chaiken & Y. Trope (Eds.), *Dual-process theories in social psychology* (pp. 383–402). The Gilford Press.

Jacoby, L. L., Shimizu, Y., Daniels, K. A., & Rhodes, M. G. (2005). Modes of cognitive control in recognition and source memory: Depth of retrieval. *Psychonomic Bulletin & Review, 12*, 852–857.

Jang, K. L., Livesley, W. J., Angleitner, A., Riemann, R., & Vernon, P. A. (2002). Genetic and environmental influences on the covariance of facets defining the domains of the five-factor model of personality. *Personality and Individual Differences, 33*(1), 83–101. Pii S0191-8869(01)00137-4.

John, O. P., & Srivastava, S. (1999). *The big-five trait taxonomy history, measurement, and theoretical perspectives* (p. 70).

Johnson, E. J., Shu, S. B., Dellaert, B. G. C., Fox, C., Goldstein, D. G., Haubl, G., et al. (2012). Beyond nudges: Tools of a choice architecture. *Marketing Letters, 23*(2), 487–504. https://doi.org/10.1007/S11002-012-9186-1

Jones, E., & Nisbett, R. A. (1972). The actor and the observer: Divergent perceptions of the cause of behavior. In E. Jones, D. Kanouse, H. Kelly, R. A. Nisbett, S. Vallins, & B. Weiner (Eds.), *Attribution: Perceiving the causes of behavior* (pp. 79–94). General Learning Press.

Joyce, R. (2001). *The myth of morality.* Cambridge University Press.

Judge, T. A., Bono, J. E., Ilies, R., & Gerhardt, M. W. (2002). Personality and leadership: A qualitative and quantitative review. *Journal of Applied Psychology, 87*(4), 765–780. https://doi.org/10.1037/0021-9010.87.4.765

Kahneman, D. (2003). A perspective on judgment and choice—Mapping bounded rationality. *American Psychologist, 9*, 697–720.

Kahneman, D. (2011). *Thinking, fast and slow.* Farrar, Straus and Giroux.

Kahneman, D., & Frederick, S. (2007). Frames and brains: Elicitation and control of response tendencies. *Trends in Cognitive Sciences, 11*(2), 45–46.

Kahneman, D., & Tversky, A. (1982). On the study of statistical intuitions. *Cognition, 11*(2), 123–141.

Kahneman, D., & Tversky, A. (2000). *Choices, values, and frames.* Cambridge University Press.

Kahneman, D., Slovic, P., & Tversky, A. (1982). *Judgment under uncertainty: Heuristics and biases.* Cambridge University Press.

Kane, R. (1996). *The significance of free will.* Oxford University Press.

Kaplan, M. (2000). To what must an epistemology be true? *Philosophy and Phenomenological Research, 61*(2), 279–304. https://doi.org/10.2307/2653652

Kauppinen, H. (2007). The rise and fall of experimental philosophy. *Philosophical Explorations, 10*, 95–118.

Keller, N., Cokely, E. T., Katsikopoulos, K., & Wegwarth, O. (2010). Naturalistic heuristics for decision making. *Journal of Cognitive Engineering and Decision Making, 4*(3), 256–274.

Kim, M., & Yuan, Y. (2015). No cross-cultural differences in the Gettier car case intuition: A replication study of Weinberg et al. 2001. *Episteme, 12*(3), 355–361. https://doi.org/10.1017/epi.2015.17

Knobe, J. (2003a). Intentional action and side effects in ordinary language. *Analysis, 63*(3), 190–194. https://doi.org/10.1111/1467-8284.00419

Knobe, J. (2003b). Intentional action in folk psychology: An experimental investigation. *Philosophical Psychology, 16*(2), 309–324. https://doi.org/10.1080/09515080320103373

Knobe, J. (2004). Intention, intentional action and moral considerations. *Analysis, 64*(2), 181–187. Retrieved from <Go to ISI>://WOS:000220663600015.

Knobe, J. (2006). The concept of intentional action: A case study in the uses of folk psychology. *Philosophical Studies, 130*(2), 203–231. https://doi.org/10.1007/S11098-004-4510-0

Knobe, J. (2010a). Action trees and moral judgment. *Topics in Cognitive Science, 2*(3), 555–578. https://doi.org/10.1111/J.1756-8765.2010.01093.X

Knobe, J. (2010b). Person as scientist, person as moralist. *Behavioral and Brain Sciences, 33*(4), 315–329; discussion 329–365. https://doi.org/10.1017/S0140525X10000907

Knobe, J. (2014). Free will and the scientific vision. In E. Machery (Ed.), *Current controversies in experimental philosophy* (pp. 69–85). Routledge.

Knobe, J., & Burra, A. (2006). The folk concepts of intention and intentional action: A cross-cultural study. *Journal of Cognition and Culture, 6*, 113–132.

Knobe, J., & Doris, J. M. (2010). Responsibility. In J. M. Doris (Ed.), *The moral psychology handbook* (pp. 321–354). Oxford University Press.

Knobe, J., & Mendlow, G. (2004). The good, the bad, and the blameworthy: Understanding the role of evaluative reasoning in folk psychology. *Journal of Theoretical and Philosophical Psychology, 24*, 252–258.

Koehler, J. J., & Gershoff, A. D. (2003). Betrayal aversion: When agents of protection become agents of harm. *Organizational Behavior and Human Decision Processes, 90*(2), 244–261. https://doi.org/10.1016/S0749-5978(02)00518-6

Koehler, J. J., & Gershoff, A. D. (2005). Betrayal aversion is reasonable. *Behavioral and Brain Sciences, 28*(4), 556–+. Retrieved from <Go to ISI>://000232624200048.

Kornblith, H. (1998). The role of intuition in philosophical inquiry: An account with no unnatural ingredients. In M. DePaul & W. Ramsey (Eds.), *Rethinking intuition: The psychology of intuition and its role in philosophical inquiry* (pp. 129–141). Rowman and Littlefield.

Kornblith, H. (2002). *Knowledge and its place in nature*. Clarendon Press; Oxford University Press.

Kuhn, D. (1991). *The skills of argument*. Cambridge University Press.

Lam, B. (2010). Are Cantonese-speakers really descriptivists? Revisiting cross-cultural semantics. *Cognition, 115*(2), 320–329. https://doi.org/10.1016/J.Cognition.2009.12.018

Langton, R. (2001). Virtues of resentment. *Metaphilosophy, 13*, 255–262.

Leistico, A. M., Salekin, R. T., DeCoster, J., & Rogers, R. (2008). A large-scale meta-analysis relating the hare measures of psychopathy to antisocial conduct.

Law and Human Behavior, 32(1), 28–45. https://doi.org/10.1007/s10979-007-9096-6

Leslie, A., Knobe, J., & Cohen, A. (2006). Acting intentionally and the side-effect effect: Theory of mind and moral judgments. *Psychological Science, 17*, 421–427.

Levi, I. (1967). *Gambling with truth; an essay on induction and the aims of science*. Knopf.

Levin, I. P., Johnson, R. D., & Davis, M. L. (1987). How information frame influences risky decisions—Between-subjects and within-subject comparisons. *Journal of Economic Psychology, 8*(1), 43–54. https://doi.org/10.1016/0167-4870(87)90005-5

Levin, I. P., Schneider, S. L., & Gaeth, G. J. (1998). All frames are not created equal: A typology and critical analysis of framing effects. *Organizational Behavior and Human Decision Processes, 76*(2), 149–188. https://doi.org/10.1006/Obhd.1998.2804

Lewis, D. (1983). *Philosophical papers*. Oxford University Press.

Livengood, J., Sytsma, J., Feltz, A., Scheines, A., & Machery, E. (2010). Philosophical temperament. *Philosophical Psychology, 23*, 313–330.

Lucas, R. E., & Fujita, F. (2000). Factors influencing the relation between extraversion and pleasant affect. *Journal of Personality and Social Psychology, 79*(6), 1039–1056. https://doi.org/10.1037//0022-3514.79.6.1039

Lucas, R. E., Diener, E., Grob, A., Suh, E. M., & Shao, L. (2000). Cross-cultural evidence for the fundamental features of extraversion. *Journal of Personality and Social Psychology, 79*(3), 452–468. https://doi.org/10.1037/0022-3514.79.3.452

Ludwig, K. (2007). The epistemology of thought examples: First vs. third person approaches. *Midwest Studies in Philosophy, 31*, 128–159.

Ludwig, K. (2010). Intuitions and relativity. *Philosophical Psychology, 23*, 427–445.

Luo, X. G., Kranzler, H. R., Zuo, L. J., Wang, S. A., & Gelernter, J. (2007). Personality traits of agreeableness and extraversion are associated with ADH4 variation. *Biological Psychiatry, 61*(5), 599–608. https://doi.org/10.1016/J.Biopsych.2006.05.017

Lycan, W. G. (1986). Moral facts and moral knowledge. *Southern Journal of Philosophy, 24*, 79–94. Retrieved from <Go to ISI>://A1986D560300006.

Lycan, W. G. (2004). Free will and the burden of proof. In A. O'Hear (Ed.), *Minds and persons: Royal Institute of Philosophy Supplement* (pp. 129–141). Cambridge University Press.

Machery, E. (2008). The folk concept of intentional action: Philosophical and experimental issues. *Mind & Language, 23*, 165–189.

Machery, E. (2017). *Philosophy within its proper bounds* (1st ed.). Oxford University Press.

Machery, E., Mallon, R., Nichols, S., & Stich, S. P. (2004). Semantics, cross-cultural style. *Cognition, 92*(3), B1–B12. https://doi.org/10.1016/J.Cognition.2003.10.003

Mackie, J. L. (1977). *Ethics: Inventing right and wrong*. Penguin.

Maher, P. (1996). Subjective and objective confirmation. *Philosophy of Science, 63*, 149–174.

Malle, B. F. (2006). Intentionality, morality, and their relationship in human judgment. *Journal of Cognition and Culture, 6*, 87–112.

Malle, B. F., & Knobe, J. (1997). Which behaviors do people explain? A basic actor-observer asymmetry. *Journal of Personality and Social Psychology, 72*(2), 288–304. Retrieved from <Go to ISI>://A1997WH68100004.

Malle, B. F., & Nelson, S. E. (2003). Judging mens rea: The tension between folk concepts and legal concepts of intentionality. *Behavioral Sciences & the Law, 21*(5), 563–580. https://doi.org/10.1002/Bsl.554

Malle, B. F., Moses, L. J., & Baldwin, D. A. (2001). *Intentions and intentionality: Foundations of social cognition*. MIT Press.

Malle, B. F., Knobe, J. A., & Nelson, S. E. (2007). Actor-observer asymmetries in explanations of behavior: New answers to an old question. *Journal of Personality and Social Psychology, 93*(4), 491–514. https://doi.org/10.1037/0022-3514.93.4.491

Mandelbaum, D., & Ripley, D. (2012). Explaining the abstract/concrete paradoxes in moral psychology: The NBAR hypothesis. *Review of Philosophy and Psychology, 3*, 351–368.

McCann, H. (1998). *The works of agency: On human action, will, and freedom*. Cornell University Press.

McCann, H. (2005). Intentional action and intending: Recent empirical studies. *Philosophical Psychology, 18*, 737–748.

McCrae, R. R. (2002). Cross-cultural research on the Five-Factor model of personality. *Online Reading in Psychology and Culture, 4*. https://doi.org/10.9707/2307-0919.1038

McCrae, R. R., & Costa, P. T. (1990). *Personality in adulthood*. Guilford Press.

McKenna, M. (2008). A hard-line reply to Pereboom's four-case Manipulation Argument (Derk Pereboom). *Philosophy and Phenomenological Research, 77*(1), 142–159. https://doi.org/10.1111/J.1933-1592.2008.00179.X

McKenzie, C. R. (2003). Rational models as theories–not standards–of behavior. *Trends in Cognitive Sciences, 7*(9), 403–406.

McNaughton, D., & Rawling, J. (2006). Deontology. In D. Copp (Ed.), *The Oxford handbook of ethical theory* (pp. 424–458). Oxford University Press.

Mcneil, B. J., Pauker, S. G., Sox, H. C., & Tversky, A. (1982). On the elicitation of preferences for alternative therapies. *New England Journal of Medicine, 306*(21), 1259–1262. https://doi.org/10.1056/Nejm198205273062103

Mele, A. (1992). *Springs of action: Understanding intentional behavior*. Oxford University Press.
Mele, A. (1995). *Autonomous agents: From self-control to autonomy*. Oxford University Press.
Mele, A. (2001). Acting intentionally: Probing folk intuitions. In B. F. Malle, L. J. Moses, & D. A. Baldwin (Eds.), *Intentions and intentionality* (pp. 27–43). MIT Press.
Mele, A. (2006). *Free will and luck*. Oxford University Press.
Mele, A. (2008). *Free will and luck*. Oxford University Press.
Mele, A., & Cushman, F. (2006). Intentional action, folk judgments, and stories: Sorting things out. *Philosophy and the Empirical, 31*, 184–201.
Merritt, A., Karlsson, L., & Cokely, E. T. (2010). Category learning and adaptive benefits of aging. In *Proceedings of the 32st Annual Conference of the Cognitive Science Society* (pp. 405–410). Cognitive Science Society.
Merz, J. F., & Fischhoff, B. (1990). Informed consent does not mean rational consent—Cognitive limitations on decision-making. *Journal of Legal Medicine, 11*(3), 321–350. Retrieved from <Go to ISI>://WOS:A1990EK30400002.
Mill, J. S., & Williams, G. (1993). *Utilitarianism; On liberty; Considerations on representative government; Remarks on Bentham's philosophy* (New ed.). Tuttle.
Miller, C. (2013). *Moral character: An empirical theory*. Oxford University Press.
Miller, J. S., & Feltz, A. (2011). Frankfurt and the folk: An experimental investigation of Frankfurt-style cases. *Consciousness and Cognition, 20*(2), 401–414. https://doi.org/10.1016/J.Concog.2010.10.015
Molenaar, P. C. M., & Campbell, C. G. (2009). The new person-specific paradigm in psychology. *Current Directions in Psychological Science, 18*(2), 112–117. https://doi.org/10.1111/J.1467-8721.2009.01619.X
Morris, M., Nisbett, R. E., & Pent, K. (1995). Causal understanding across domains and cultures. In D. Sperber, D. Premack, & A. J. Premack (Eds.), *Causal cognition: A multidisciplinary debate* (pp. 577–612). Oxford University Press.
Mortensen, K., & Nagel, J. (2016). Armchair-friendly experimental philosophy. In J. Sytsma & W. Buckwalter (Eds.), *A companion to experimental philosophy* (pp. 53–70). Wiley-Blackwell.
Moshman, D. (2000). Diversity in reasoning and rationality: Metacognitive and developmental considerations. *Educational Psychology Papers and Publications, 46*, 689–690.
Moxley, J. H., Ericsson, K. A., Charness, N., & Krampe, R. T. (2012). The role of intuition and deliberative thinking in experts' superior tactical decision-making. *Cognition, 124*(1), 72–78.
Muenscher, R., Vetter, M., & Scheuerle, T. (2016). A review and taxonomy of choice architecture techniques. *Journal of Behavioral Decision Making, 29*(5), 511–524. https://doi.org/10.1002/bdm.1897

Nadelhoffer, T. (2004). On praise, side effects, and folk ascriptions of intentionality. *Journal of Theoretical and Philosophical Psychology, 24*, 196–213.

Nadelhoffer, T. (2006a). Bad acts, blameworthy agents, and intentional actions: Some problems for jury impartiality. *Philosophical Explorations, 9*, 203–220.

Nadelhoffer, T. (2006b). On trying to save the simple view. *Mind & Language, 21*(5), 565–586. https://doi.org/10.1111/J.1468-0017.2006.00292.X

Nadelhoffer, T., & Feltz, A. (2008). The actor-observer bias and moral intuitions: Adding fuel to Sinnott-Armstrong's fire. *Neuroethics, 1*, 133–144.

Nadelhoffer, T., Kvaran, T., & Nahmias, E. (2009). Temperament and intuition: A commentary on Feltz and Cokely. *Consciousness and Cognition, 18*(1), 351–355; discussion 356–358. https://doi.org/10.1016/j.concog.2008.11.006

Nadelhoffer, T., Rose, D., Buckwalter, W., & Nichols, S. (2020a). Natural compatibilism, indeterminism, and intrusive metaphysics. *Cognitive Science, 44*(8), e12873. https://doi.org/10.1111/cogs.12873

Nadelhoffer, T., Shepard, J., Crone, D. L., Everett, J. A., Earp, B. D., & Levy, N. (2020b). Does encouraging a belief in determinism increase cheating? Reconsidering the value of believing in free will. *Cognition, 203*, 104342.

Nagel, J. (2012). Mindreading in Gettier cases and skeptical pressure cases. In J. Brown & M. Gerken (Eds.), *Knowledge ascription* (pp. 171–191). Oxford University Press.

Nagel, J., Juan, V. S., & Mar, R. A. (2013). Lay denial of knowledge for justified true beliefs. *Cognition, 129*(3), 652–661. https://doi.org/10.1016/j.cognition.2013.02.008

Nahmias, E. (2007). Autonomous agency and social psychology. In M. Marraffa, M. Cardero, & F. Ferretti (Eds.), *Cartographies of the mind: Philosophy and psychology in intersection* (pp. 169–185). Springer.

Nahmias, E., & Murray, D. (2010). Experimental philosophy on free will: An error theory for incompatibilist intuitions. In J. A. Aguilar, A. Buckareff, & K. Frankish (Eds.), *New waves in philosophy of action* (pp. 189–215). Palgrave Macmillan.

Nahmias, E., Morris, S., Nadelhoffer, T., & Turner, J. (2005). Surveying freedom: Folk intuitions about free will and moral responsibility. *Philosophical Psychology, 18*(5), 561–584. https://doi.org/10.1080/09515080500264180

Nahmias, E., Coates, D., & Kvaran, T. (2006a). Free will, moral responsibility, and mechanism: Experiments on folk intuitions. *Philosophy and the Empirical, 31*, 214–242.

Nahmias, E., Morris, S., Nadelhoffer, T., & Turner, J. (2006b). Is incompatibilism intuitive? *Philosophy and Phenomenological Research, 73*(1), 28–53. https://doi.org/10.1111/J.1933-1592.2006.Tb00603.X

Newell, A. (1973). You can't play 20 questions with nature and win: Projective comments on the papers of this symposium. In W. G. Chase (Ed.), *Visual information processing*. Academic Press.

Newstead, S. E. (2000). Are there two different types of thinking? *Behavioral & Brain Sciences, 23*, 690–691.

Nichols, S. (2004a). After objectivity: An empirical study of moral judgment. *Philosophical Psychology, 17*(1), 3–26. https://doi.org/10.1080/0951508042000202354

Nichols, S. (2004b). The folk psychology of free will: Fits and starts. *Mind & Language, 19*(5), 473–502. https://doi.org/10.1111/J.0268-1064.2004.00269.X

Nichols, S. (2007). The rise of compatibilism: A case study in the quantitative history of philosophy. *Midwest Studies in Philosophy, 31*, 260–270.

Nichols, S., & Knobe, J. (2007). Moral responsibility and determinism: The cognitive science of folk intuitions. *Nous, 41*(4), 663–685. https://doi.org/10.1111/J.1468-0068.2007.00666.X

Nichols, S., & Ulatowski, J. (2007). Intuitions and individual differences: The Knobe effect revisited. *Mind & Language, 22*(4), 346–365. https://doi.org/10.1111/J.1468-0017.2007.00312.X

Nisbett, R. E., & Wilson, T. D. (1977). Telling more than we can know: Verbal reports on mental processes. *Psychological Review, 84*(3), 231–259. https://doi.org/10.1037/0033-295x.84.3.231

Nisbett, R. E., Legant, P., & Marecek, J. (1973). Behavior as seen by actor and as seen by observer. *Journal of Personality and Social Psychology, 27*(2), 154–164. https://doi.org/10.1037/H0034779

Nisbett, R. E., Peng, K. P., Choi, I., & Norenzayan, A. (2011). Culture and systems of thought: Comparison of holistic and analytic cognition. *Psikhologicheskii Zhurnal, 32*(1), 55–86.

Oakley, J. (1996). Varieties of Virtue Ethics. *Ratio-New Series, 9*(2), 128–152. https://doi.org/10.1111/J.1467-9329.1996.Tb00101.X

Okan, Y., Garcia-Retamero, R., Cokely, E. T., & Maldonado, A. (2012a). Individual differences in graph literacy: Overcoming denominator neglect in risk comprehension. *Journal of Behavioral Decision Making, 25*(4), 390–401.

Okan, Y., Garcia-Retamero, R., Galesic, M., & Cokely, E. T. (2012b). When higher bars are not larger quantities: On individual differences in the use of spatial information in graph comprehension. *Spatial Cognition and Computation, 12*(2–3), 195–218.

Okan, Y., Garcia-Retamero, R., Cokely, E. T., & Maldonado, A. (2015). Improving risk understanding across ability levels: Encouraging active processing with dynamic icon arrays. *Journal of Experimental Psychology: Applied, 21*(2), 178–194.

Okan, Y., Garcia-Retamero, R., Cokely, E. T., & Maldonado, A. (2018). Biasing and debiasing health decisions with bar graphs: Costs and benefits of graph literacy. *Quarterly Journal of Experimental Psychology, 71*(12), 2506–2519.

Oliver, A. (2015). Nudging, shoving, and budging: Behavioural economic-informed policy. *Public Administration, 93*, 700–714.

Open Science Collaboration. (2015). PSYCHOLOGY. Estimating the reproducibility of psychological science. *Science, 349*(6251), aac4716. https://doi.org/10.1126/science.aac4716

Osbeck, L. (2001). Direct apprehension and social construction: Revisiting the concept of intuition. *Journal of Theoretical and Philosophical Psychology, 21*, 118–131.

Osman, M. (2004). An evaluation of dual-process theories of reasoning. *Psychonomic Bulletin & Review, 11*(6), 988–1010.

Ostendorf, F., & Angleitner, A. (2004). NEO-Persönlichkeitsinventar nach Costa und McCrae: NEO-PI-R; Manual Revidierte Fassung. Göttingen: Hogrefe.

Pachur, T., & Galesic, M. (2013). Strategy selection in risky choice: The impact of numeracy, affect, and cross-cultural differences. *Journal of Behavioral Decision Making, 26*(3), 260–271.

Patrick, C. J. (2006). *Handbook of psychopathy*. Guilford Press.

Payne, J. W., Bettman, J. R., Coupey, E., & Johnson, E. J. (1992). A constructive process view of decision making: Multiple strategies in judgment and choice. *Acta Psychologica, 80*(1–3), 107–141. https://doi.org/10.1016/0001-6918(92)90043-D

Payne, J. W., Bettman, J. R., & Johnson, E. J. (1993). *The adaptive decision maker*. Cambridge University Press. https://doi.org/10.1017/CBO9781139173933

Pereboom, D. (1995). Determinism Al-Dente + hard-determinism. *Nous, 29*(1), 21–45. https://doi.org/10.2307/2215725

Pereboom, D. (2001). *Living without free will*. Cambridge University Press.

Peters, E. (2012). Beyond comprehension the role of numeracy in judgments and decisions. *Current Directions in Psychological Science, 21*(1), 31–35.

Peters, E., Västfjäll, D., Slovic, P., Mertz, C. K., Mazzocco, K., & Dickert, S. (2006). Numeracy and decision making. *Psychological Science, 17*, 407–413.

Peterson, C., & Seligman, M. E. P. (2004). *Character strengths and virtues: A handbook and classification*. American Psychological Association; Oxford University Press.

Petrinovich, L., & O'Neill, P. (1996). Influence of wording and framing effects on moral intuitions. *Ethology and Sociobiology, 17*(3), 145–171. https://doi.org/10.1016/0162-3095(96)00041-6

Petrova, D., Garcia-Retamero, R., & Cokely, E. T. (2015). Understanding the harms and benefits of cancer screening: A model of factors that shape informed decision making. *Medical Decision Making, 35*(7), 847–858.

Petrova, D., Garcia-Retamero, R., Catena, A., Cokely, E., Carrasco, A. H., Moreno, A. A., & Hernández, J. A. R. (2017). Numeracy predicts risk of pre-hospital decision delay: A retrospective study of acute coronary syndrome survival. *Annals of Behavioral Medicine, 51*(2), 292–306.

Petrova, D., Kostopoulou, O., Delaney, B. C., Cokely, E. T., & Garcia-Retamero, R. (2018). Strengths and gaps in physicians' risk communication: A scenario study of the influence of numeracy on cancer screening communication. *Medical Decision Making, 38*(3), 355–365.

Petrova, D., Cokely, E. T., Sobkow, A., Traczyk, J., Garrido, D., & Garcia-Retamero, R. (2023). Measuring feelings about choices and risks: The Berlin Emotional Responses to Risk Instrument (BERRI). *Risk Analysis, 43*, 724–746.

Pettit, D., & Knobe, J. (2009). The pervasive impact of moral judgment. *Mind & Language, 24*, 586–604.

Petushek, E. J., Cokely, E. T., Ward, P., Krosshaug, T., & Myer, G. (2014). Visual assessment of ACL injury risk: Can expertise be achieved? *British Journal of Sports Medicine, 48*(7), 652–652.

Petushek, E. J., Cokely, E. T., Ward, P., Durocher, J. J., Wallace, S. J., & Myer, G. D. (2015a). Injury risk estimation expertise: Assessing the ACL injury risk estimation quiz. *The American Journal of Sports Medicine, 43*(7), 1640–1647.

Petushek, E. J., Cokely, E. T., Ward, P., & Myer, G. D. (2015b). Injury risk estimation expertise: Cognitive-perceptual mechanisms of ACL-IQ. *Journal of Sport and Exercise Psychology, 37*(3), 291–304.

Phelan, M., & Sarkissian, H. (2008). The folk strike back: Or, why you didn't do it intentionally, though it was bad and you knew it. *Philosophical Studies, 138*, 291–298.

Pink, T. (2004). *Free will: A very short introduction.* Oxford University Press.

Prinz, J. J. (2007). *The emotional construction of morals.* Oxford University Press.

Pust, J. (2000). *Intuitions as evidence.* Garland Pub.

Pust, J. (2001). Against explanationist skepticism regarding philosophical intuitions. *Philosophical Studies, 106*, 227–258.

Ramasubramanian, M., Allan, J. N., Garcia-Retamero, R., Jenkins-Smith, H., & Cokely, E. T. (2019, November). Flood risk literacy: Communication and implications for protective action. In *Proceedings of the Human Factors and Ergonomics Society Annual Meeting* (Vol. 63, No. 1, pp. 1629–1633). SAGE Publications.

Raza, M. A., Salehi, S., Ghazal, S., Ybarra, V. T., Naqvi, S. A. M., Cokely, E. T., & Teodoriu, C. (2019). Situational awareness measurement in a simulation-based training framework for offshore well control operations. *Journal of Loss Prevention in the Process Industries, 62*, 103921.

Raza, M., Kiran, R., Ghazal, S., Jeon, J., Salehi, S., Kang, Z., & Cokely, E. T. (2023). An eye tracking based framework for safety improvement of offshore operations. *Journal of Eye Movement Research, 16*.

Revelle, W., & Scherer, K. (2009). Personality and emotion. In D. Sander & K. Scherer (Eds.), *The Oxford companion to emotion and the affective sciences* (pp. 304–305). Oxford University Press.

Reyna, V. F. (2004). How people make decisions that involve risk. *Current Directions in Psychological Science, 13*, 60–66.

Reyna, V. F., Nelson, W. L., Han, P. K., & Dieckmann, N. F. (2009). How numeracy influences risk comprehension and medical decision making. *Psychological Bulletin, 135*, 943–973.

Richards, N. (1988). Is humility a virtue. *American Philosophical Quarterly, 25*(3), 253–259. Retrieved from <Go to ISI>://A1988P302800005.

Rini, R. (2014). Analogies, moral intuitions, and the expertise defense. *Review of Philosophy and Psychology, 5*, 169–181.

Rini, R. (2015). How not to test for philosophical expertise. *Synthese, 192*(2), 431–452.

Rini, R. (2016). Debuking debunking: A regress challenge for psychological threats to moral judgments. *Philosophical Studies: An International Journal for Philosophy in the Analytic Tradition, 173*(3), 675–697. Retrieved from http://www.jstor.org/stable/24703868

Roozenbeek, J., & Van der Linden, S. (2024). *The psychology of misinformation*. Cambridge University Press.

Rose, D., & Nichols, S. (2013). The lesson of bypassing. *Review of Philosophy and Psychology, 4*, 599–619.

Rose, D., Livengood, J., Sytsma, J., & Machery, E. (2012). Deep trouble for the deep self. *Philosophical Psychology, 25*, 629–646.

Ross, W. D. (1988). *The right and the good*. Hackett Pub. Co.

Rothman, A. J., & Salovey, P. (1997). Shaping perceptions to motivate healthy behavior: The role of message framing. *Psychological Bulletin, 121*(1), 3–19. https://doi.org/10.1037//0033-2909.121.1.3

Rusting, C. L., & Larsen, R. J. (1997). Extraversion, neuroticism, and susceptibility to positive and negative affect: A test of two theoretical models. *Personality and Individual Differences, 22*(5), 607–612. https://doi.org/10.1016/S0191-8869(96)00246-2

Ryan, R. M., & Deci, E. L. (2017). *Self-determination theory: Basic psychological needs in motivation, development, and wellness*. Guilford Press.

Ryff, C. (1995). Psychological well-being in adult life. *Current Directions in Psychological Science, 4*, 99–104.

Ryle, G. (1957). On forgetting the difference between right and wrong. In A. I. Melden (Ed.), *Essays in moral philosophy*. University of Washington Press.

Salehi, S., Kiran, R., Jeon, J., Kang, Z., Cokely, E. T., & Ybarra, V. T. (2018). Developing a cross-disciplinary, scenario-based training approach integrated with eye tracking data collection to enhance situational awareness in offshore

oil and gas operations. *Journal of Loss Prevention in the Process Industries, 56*, 78–94.
Sandman, L., & Munthe, C. (2010). Shared decision making, paternalism and patient choice. *Health Care Analysis, 18*(1), 60–84. https://doi.org/10.1007/S10728-008-0108-6
Sarkissian, H., Chatterjee, A., De Brigard, F., Knobe, J., Nichols, S., & Sirker, S. (2010). Is belief in free will a cultural universal? *Mind & Language, 25*(3), 346–358.
Sarkissian, H., Park, J., Tien, D., Wright, J. C., & Knobe, J. (2011). Folk moral relativism. *Mind & Language, 26*(4), 482–505. https://doi.org/10.1111/J.1468-0017.2011.01428.X
Saroglou, V. (2002). Religion and the five factors of personality: A meta-analytic review. *Personality and Individual Differences, 32*, 15–25.
Schmitt, D. P., Allik, J., McCrae, R. R., Benet-Martinez, V., Alcalay, L., Ault, L., et al. (2007). The geographic distribution of big five personality traits—Patterns and profiles of human self-description across 56 nations. *Journal of Cross-Cultural Psychology, 38*(2), 173–212. https://doi.org/10.1177/0022022106297299
Schueler, G. F. (1997). Why modesty is a virtue. *Ethics, 107*(3), 467–485. https://doi.org/10.1086/233745
Schulz, E., Cokely, E. T., & Feltz, A. (2011). Persistent bias in expert judgments about free will and moral responsibility: A test of the expertise defense. *Consciousness and Cognition, 20*(4), 1722–1731. https://doi.org/10.1016/j.concog.2011.04.007
Schwartz, S. H. (1992). Universals in the content and structure of values: Theory and empirical tests in 20 countries. In M. Zanna (Ed.), *Advances in experimental social psychology* (Vol. 25, pp. 1–65). Academic Press. https://doi.org/10.1016/S0065-2601(08)60281-6
Schwartz, S., & Bilsky, W. (1987). Toward a universal psychological structure of human values. *Journal of Personality and Social Psychology, 53*, 550–562.
Schwitzgebel, E., & Cushman, F. (2012). Expertise in moral reasoning? Order effects on moral judgment in professional philosophers and non-philosophers. *Mind & Language, 27*(2), 135–153. https://doi.org/10.1111/J.1468-0017.2012.01438.X
Schwitzgebel, E., & Rust, J. (2009). The moral behaviour of ethicists: Peer opinion. *Mind, 118*(472), 1043–1059. https://doi.org/10.1093/mind/fzp108
Schwitzgebel, E., & Rust, J. (2014). The moral behavior of ethics professors: Relationships among self-reported behavior, expressed normative attitude, and directly observed behavior. *Philosophical Psychology, 27*(3), 293–327. https://doi.org/10.1080/09515089.2012.727135
Searle, J. (1980). Minds, brains, and programs. *Behavioral and Brain Sciences, 3*, 242–417.

Seligman, M. E. P., & Csikszentmihayli, M. (2000). Positive psychology: An introduction. *American Psychologist, 55*, 5–14.

Selinger, E., & Whyte, K. (2011). Is there a right way to nudge? The practice and ethics of choice architecture. *Sociology Compass, 5*, 923–935.

Seyedsayamdost, H. (2015). On normativity and epistemic intuitions: Failure of replication. *Episteme, 12*, 95–116. https://doi.org/10.1017/epi.2014.27

Shafer-Landau, R. (2003). *Moral realism: A defence*. Clarendon; Oxford University Press.

Shafir, E., & Tversky, A. (1995). Decision making. In E. E. Smith & D. N. Osherson (Eds.), *Thinking: An invitation to cognitive science* (Vol. 3, 2nd ed., pp. 77–100). MIT Press.

Shanteau, J. (1992). Competence in experts: The role of task characteristics. *Organizational Behavior and Human Decision Processes, 53*(2), 252–266. https://doi.org/10.1016/0749-5978(92)90064-E

Shanteau, J. (1988). Psychological characteristics and strategies of expert decision makers. *Acta Psychologica, 68*(1–3), 203–215. https://doi.org/10.1016/0001-6918(88)90056-X

Shepard, J. (2011). *Normative judgments, deep self judgments, and intentional action*. (M.A.). Georgia State University.

Shiffrin, R. M., & Schneider, W. (1977). Controlled and automatic human information processing: II. Perceptual learning, automatic attending and a general theory. *Psychological Review, 84*(2), 127.

Sibley, C. G., & Duckitt, J. (2008). Personality and prejudice: A meta-analysis and theoretical review. *Personality and Social Psychology Review, 12*(3), 248–279. https://doi.org/10.1177/1088868308319226

Simon, H. (1955). A behavioral model of rational choice. *Quarterly Journal of Economics, 69*, 99–118.

Simon, H. (1990). Invariants of human behavior. *Annual Review of Psychology, 41*, 1–19.

Singer, P. (1972). Moral experts. *Analysis, 32*, 115–117.

Singer, P. (1993). *Practical ethics* (2nd ed.). Cambridge University Press.

Sinnott-Armstrong, W. (2008). Abstract + concrete = paradox? In J. Knobe & S. Nichols (Eds.), *Experimental philosophy* (pp. 209–230). Oxford University Press.

Skinner, B. F. (1971). *Beyond freedom and dignity* (1st ed.). Knopf.

Slote, M. A. (2001). *Morals from motives*. Oxford University Press.

Smilansky, S. (2000). *Free will and illusion*. Oxford University Press.

Smilansky, S. (2002). Free will, fundamental dualism, and the centrality of illusion. In R. Kane (Ed.), *The Oxford handbook of free will* (pp. 489–505). Oxford University Press.

Smith, M. (1995). *The moral problem*. Blackwell.

Smith, N., Goldstein, D. G., & Johnson, E. J. (2013). Choice without awareness: Ethical and policy implications of defaults. *Journal of Public Policy & Marketing, 32*, 159–172.

Soll, J., Milkman, K., & Payne, J. W. (2015). A user's guide to debiasing. In G. Keren & G. Wu (Eds.), *Wiley-Blackwell handbook of judgment and decision making* (pp. 924–951). Wiley.

Sommers, T. (2010). Experimental philosophy and free will. *Philosophy Compass, 5*, 192–212.

Sommers, T. (2012). *Relative justice: Cultural diversity, free will, and moral responsibility*. Princeton University Press.

Sommers, T. (2014). Free will and experimental philosophy: An intervention. In J. Clausen & N. Levy (Eds.), *Handbook of neuroethics* (pp. 273–286). Springer.

Sosa, E. (2007a). Experimental philosophy and philosophical intuition. *Philosophical Studies, 132*(1), 99–107. https://doi.org/10.1007/S11098-006-9050-3

Sosa, E. (2007b). Intuitions: Their nature and epistemic efficacy. *Grazer Philosophische Studien, 74*, 51–67.

Sosa, E. (2009). A defense of the use of intuitions in philosophy. In M. Bishop & D. Murphy (Eds.), *Stich and his critics* (pp. 101–112). Wiley.

Spinath, F., & Johnson, W. (2011). Behavior genetics. In T. Chamorro-Premuzic, S. von Strumm, & A. Furnham (Eds.), *The Wiley-Blackwell handbook of individual differences* (pp. 271–304). Blackwell Publishing.

Sripada, C. (2010). The deep self model and asymmetries in folk judgments about intentional actions. *Philosophical Studies, 151*, 159–176.

Sripada, C. S. (2012). What makes a manipulated agent unfree? *Philosophy and Phenomenological Research, 85*(3), 563–593. https://doi.org/10.1111/J.1933-1592.2011.00527.X

Stanovich, K. E. (1999). *Who is rational?: Studies of individual differences in reasoning*. Lawrence Erlbaum Associates.

Stanovich, K. E., & West, R. F. (1998). Individual differences in rational thought. *Journal of Experimental Psychology: General, 127*, 161–188.

Stanovich, K. E., & West, R. F. (2000). Advancing the rationality debate. *Behavioral and Brain Sciences, 23*(05), 701–717.

Stanovich, K. E., & West, R. F. (2008). On the relative independence of thinking biases and cognitive ability. *Journal of Personality and Social Psychology, 94*, 672–695.

Stich, S. (1990). *The fragmentation of reason*. MIT Press.

Stich, S. (1998). Reflective equilibrium, analytic epistemology and the problem of cognitive diversity. *Synthese, 74*(3), 391–413. https://doi.org/10.1007/Bf00869637

Stich, S., & Tobia, K. (2016). Experimental philosophy's challenge to the "Great Tradition". *Analytica: Revista de Filosofia, 20*, 9–40.

Stillman, T. F., Baumeister, R. F., & Mele, A. R. (2011). Free will in everyday life: Autobiographical accounts of free and unfree actions. *Philosophical Psychology, 24*(3), 381–394. https://doi.org/10.1080/09515089.2011.556607

Strawson, G. (1986). *Freedom and belief.* Clarendon Press; Oxford University Press.

Strawson, G. (1994). The impossibility of moral responsibility. *Philosophical Studies, 75*(1–2), 5–24. https://doi.org/10.1007/Bf00989879

Sunstein, C. R. (2005). Moral heuristics. *Behavioral and Brain Sciences, 28*(4), 531–+. Retrieved from <Go to ISI>://000232624200032.

Sunstein, C. R., & Thaler, R. H. (2003). Libertarian paternalism is not an oxymoron. *University of Chicago Law Review, 70*(4), 1159–1202. https://doi.org/10.2307/1600573

Sunstein, C. R., Reisch, L. A., & Kaiser, M. (2019). Trusting nudges? Lessons from an international survey. *Journal of European Public Policy, 26*(10), 1417–1443. https://doi.org/10.1080/13501763.2018.1531912

Sverdlik, S. (2004). Intentionality and moral judgments in commonsense thought about action. *Journal of Theoretical and Philosophical Psychology, 24*, 224–236.

Swain, S., Alexander, J., & Weinberg, J. (2008). The instability of philosophical intuitions: Running hot and cold on Truetemp (Keith Lehrer). *Philosophy and Phenomenological Research, 76*(1), 138–155.

Swanton, C. (2003). *Virtue ethics: A pluralistic view*. Oxford University Press.

Swanton, C. (2010). A challenge to rational virtue from moral virtue: The case of universal love. *Metaphilosophy, 41*, 152–171.

Tangney, J. P., Baumeister, R. F., & Boone, A. L. (2004). High self-control predicts good adjustment, less pathology, better grades, and interpersonal success. *Journal of Personality, 72*(2), 271–324. https://doi.org/10.1111/J.0022-3506.2004.00263.X

Thaler, R., & Sunstein, C. (2003). Libertarian paternalism. *American Economic Review, 93*(2), 175–179. https://doi.org/10.1257/000282803321947001

Thaler, R., & Sunstein, C. (2008). *Nudge: Improving decisions about health, wealth, and happiness.* Yale University Press.

Thaler, R. H., Sunstein, C. R., & Balz, J. (2012). Choice architecture. In E. Shafir (Ed.), *The behavioral foundations of public policy* (pp. 428–439). Princeton University Press.

Tobia, K., Buckwalter, W., & Stich, S. (2013). Moral intuitions: Are philosophers experts? *Philosophical Psychology, 26*(5), 629–638. https://doi.org/10.1080/09515089.2012.696327

Toplak, M. E., West, R. F., & Stanovich, K. E. (2011). The Cognitive Reflection Test as a predictor of performance on heuristics-andbiases tasks. *Mem Cognit., 39*(7), 1275–89. https://doi.org/10.3758/s13421-011-0104-1. PMID: 21541821

Trout, J. D. (2005). Paternalism and cognitive bias. *Law and Philosophy, 137*, 926–931.

Turilli, M., & Floridi, L. (2009). The ethics of information transparency. *Ethics and Information Technology, 11*(2), 105–112. https://doi.org/10.1007/s10676-009-9187-9

Turner, M. L., & Engle, R. W. (1989). Is working memory capacity task dependent. *Journal of Memory and Language, 28*(2), 127–154. https://doi.org/10.1016/0749-596x(89)90040-5

Turner, J., & Nahmias, E. (2006). Are the folk agent-causationists? *Mind & Language, 21*(5), 597–609. https://doi.org/10.1111/J.1468-0017.2006.00295.X

Tversky, A., & Kahneman, D. (1974). Judgment under uncertainty: Heuristics and biases. *Science, 185*(4157), 1124–1131.

Tversky, A., & Kahneman, D. (1981). The framing of decisions and the psychology of choice. *Science, 211*(4481), 453–458. https://doi.org/10.1126/Science.7455683

Van der Linden, S. (2023). *Foolproof: Why misinformation infects our minds and how to build immunity.* WW Norton & Company.

Van Inwagen, P. (1983). *An essay on free will.* Clarendon Press; Oxford University Press.

Van Inwagen, P. (1997). Materialism and the psychological continuity account of personal identity. *Philosophy and Phenomenological Research, 31*, 305–319.

Vargas, M. (2005). The revisionist's guide to responsibility. *Philosophical Studies, 125*(3), 399–429. https://doi.org/10.1007/S11098-005-7783-Z

Vargas, M. (2006). Philosophy and the folk: On some implications of experimental work for philosophical debates on free will. *The Journal of Cognition and Culture, 6*, 249–264.

Velleman, J. D. (2008). A theory of value. *Ethics, 118*, 410–436.

Vohs, K. D., & Schooler, J. W. (2008). The value of believing in free will—Encouraging a belief in determinism increases cheating. *Psychological Science, 19*(1), 49–54. https://doi.org/10.1111/J.1467-9280.2008.02045.X

Wallace, R. J. (1994). *Responsibility and the moral sentiments.* Harvard University Press.

Watson, G. (1975). Free agency. *Journal of Philosophy, 72*(8), 205–220. https://doi.org/10.2307/2024703

Weimer, D., & Vining, A. (2017). *Policy Analysis: Concepts and Practice* (6th ed.). Routledge. https://doi.org/10.4324/9781315442129

Weinberg, J. (2006). How to challenge intuitions empirically without risking skepticism. *Philosophy and the Empirical, 31*, 318–343.

Weinberg, J., & Alexander, J. (2014). The challenge of sticking with intuitions through thick and thin. https://doi.org/10.1093/acprof:oso/9780199609192.003.0011.

Weinberg, J., & Crowley, S. (2009). Loose constituivity and armchair philosophy. *Studie Philosophia Estonica, 23*, 332–355.

Weinberg, J., Nichols, S., & Stich, S. P. (2001). Normativity and epistemic intuitions. *Philosophical Topics, 29*, 429–460.

Weinberg, J. M., Gonnerman, C., Buckner, C., & Alexander, J. (2010). Are philosophers expert intuiters? *Philosophical Psychology, 23*(3), 331–355. https://doi.org/10.1080/09515089.2010.490944

Weinberg, J. M., Alexander, J., Gonnerman, C., & Reuter, S. (2012a). Restrictionism and reflection: Challenge deflected, or simply redirected? *Monist, 95*(2), 200–222. https://doi.org/10.5840/monist201295212

Weinberg, J., Crowley, S., Gonnerman, C., Vandewalker, I., & Swain, S. (2012b). Intuition & calibration. *Essays in Philosophy, 13*, 257–284. https://doi.org/10.5840/eip201213115

Weirich, P. (2004). *Realistic decision theory: Rules for nonideal agents in nonideal circumstances.* Oxford University Press.

Welch, B. F. (2013). Shifting the concept of nudge. *The Journal of Medical Ethics, 39*(8), 497–498. https://doi.org/10.1136/medethics-2012-101111

Wickens, C. D., Hollands, J. G., Banbury, S., & Parasuraman, R. (2013). *Engineering psychology and human performance* (4th ed.). Pearson.

Williamson, T. (2007). *The philosophy of philosophy.* Blackwell Pub.

Williamson, T. (2011). Philosophical expertise and the burden of proof. *Metaphilosophy, 42*(3), 215–229.

Wilt, J., & Revelle, W. (2009). Extraversion. In M. Leary & R. Hoyle (Eds.), *Handbook of individual differences in social behavior* (pp. 27–45). Guilford Press.

Winkler, B. (2000). *Which kind of transparency? On the need for clarity in monetary policy making.* European Central Bank.

Wolf, S. R. (1990). *Freedom within reason.* Oxford University Press.

Woller-Carter, M. M., Okan, Y., Cokely, E. T., & Garcia-Retamero, R. (2012). Communicating and distorting risks with graphs: An eye-tracking study. In *Proceedings of the Human Factors and Ergonomics Society Annual Meeting* (Vol. 56, No. 1, pp. 1723–1727). SAGE Publications.

Wong, T. J., Cokely, E. T., & Schooler, L. J. (2010). An online database of ACT-R parameters: Towards a transparent community-based approach to model development. In *Proceedings of the 13th International Conference on Cognitive Modeling* (pp. 282–286). Drexel University.

Wright, C. (1992). *Truth and objectivity.* Harvard University Press.

Wright, J. C., Cullum, J., & Schwab, N. (2008). The cognitive and affective dimensions of moral conviction: Implications for attitudinal and behavioral measures of interpersonal tolerance. *Personality and Social Psychology Bulletin, 34*(11), 1461–1476. https://doi.org/10.1177/0146167208322557

Ybarra, V., Cokely, E. T., Adams, C., Woller-Carter, M., Allan, J., Feltz, A., & Garcia-Retamero, R. (2017). Training graph literacy: Developing the risk literacy. org Outreach Platform. In *CogSci.*

Zagzebski, L. (2010). Exemplarist virtue theory. *Metaphilosophy, 41*(1–2), 41–57. Retrieved from <Go to ISI>://000273317000003.

Zelenski, J. M., & Larsen, R. J. (2002). Predicting the future: How affect-related personality traits influence likelihood judgments of future events. *Personality and Social Psychology Bulletin, 28*(7), 1000–1010. https://doi.org/10.1177/01467202028007012

Ziółkowski, A., Wiegmann, A., Horvath, J., et al. (2023). Truetemp cooled down: The stability of Truetemp intuitions. *Synthese, 201*, 108. https://doi.org/10.1007/s11229-023-04055-z

Index[1]

[1] Note: Page numbers followed by 'n' refer to notes.

A. Feltz, E. T. Cokely, *Diversity and Disagreement*,
https://doi.org/10.1007/978-3-031-61935-9

F

Printed in the USA
CPSIA information can be obtained
at www.ICGtesting.com
CBHW072124210724
11924CB00005B/360

9 783031 619342